THE PRECIOUS METALS OF MEDICINE

By the same authors

THE MEDICAL GARDEN

WOMEN IN WHITE

THE STORY OF MEDICINE IN AMERICA

The Precious Metals of Medicine

by

GEOFFREY MARKS

and

WILLIAM K. BEATTY

Illustrated with photographs

CHARLES SCRIBNER'S SONS

New York

Library of Congress Cataloging in Publication Data

Marks, Geoffrey.
The precious metals of medicine.

Bibliography: p.
1. Metals—Therapeutic use—History. 2. Metals in surgery—History. I. Beatty, William K., 1926– joint author. II. Title. [DNLM: 1. Metals—Therapeutic use—Popular works. QV290 M346p]
RM301.M37 610'.28 74-12333
ISBN 0-684-13979-0

This book published simultaneously in the United States of America and in Canada—Copyright under the Berne Convention

1 3 5 7 9 11 13 15 17 19 V/C 20 18 16 14 12 10 8 6 4 2

Printed in the United States of America

for

John W. Rippon

and

Philip C. Hoffmann

who are

lovely and pleasant in their lives

Contents

Preface

Metals have played a variety of roles in the history of medicine. In the beginning, their use was superstitious and empirical; then crude surgery was performed with crude metal instruments. But in the forward march of diagnosis, therapy, instrumentation, and surgery toward what we now regard as "modern" medicine, metals continued to make their contribution.

Many men, famous and obscure, labored to prove or defend the efficacy, sometimes ephemeral, of metal medicaments, and there were colorful figures among them. The legendary Basil Valentine rode his chariot of antimony triumphantly; the rebellious Paracelsus, who separated the chemists from the alchemists, used metals as a stepping-stone to scientific medicine; Elisha Perkins, an impoverished general practitioner, achieved (shortlived) fame and made a considerable amount of money with his "metallic tractors"; Henry Wicker, in search of water for his cattle during a drought, discovered Epsom salt.

This book offers a panoramic view of the role of metals in medicine, exemplifying their manifold uses. Some of these uses have bowed to increased scientific knowledge; others have flowered only in recent years. The field is so large a one that we have felt impelled to limit the text to the medical uses of metals, excluding the many applications of metals to dentistry and other health sciences. Also, because attention is focused on the *precious* aspects of metals in medical performance, discussion of the poisonous effects and occupational hazards has not generally been included.

A Portfolio of Pictures

Votive tablet found on the site of the Temple of Aesculapius in Athens, showing a box of scalpels and two cupping glasses

The portrait of Basil Valentine *(opposite)* in the Munich Royal Cabinet of Etchings bears a striking resemblance to the sixteenth-century Swiss physician Paracelsus *(above)*, who is believed by many to have been the true author of the work entitled *Basil Valentine His Triumphant Chariot of Antimony,* first published in 1604

BASIL VALENTINE
HIS
Triumphant Chariot
OF
ANTIMONY,
WITH
ANNOTATIONS
OF
Theodore Kirkringius. M. D.
WITH

The True Book of the Learned *Synesius* a Greek Abbot taken out of the Emperour's Library, concerning the Philosopher's Stone.

LONDON.

Printed for *Dorman Newman* at the Kings Arms in the *Poultry*. 1678.

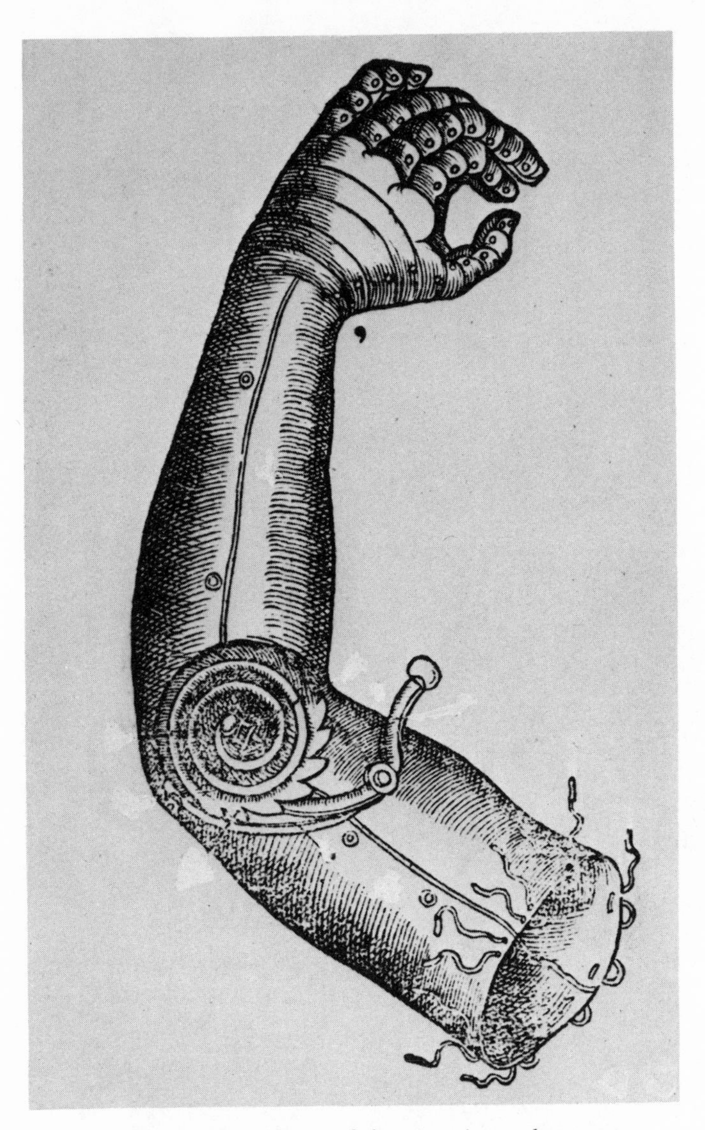

Metal arm prosthesis designed by the sixteenth-century French surgeon Ambroise Paré

Opposite: Title page of one of the many editions of the *Triumphant Chariot*

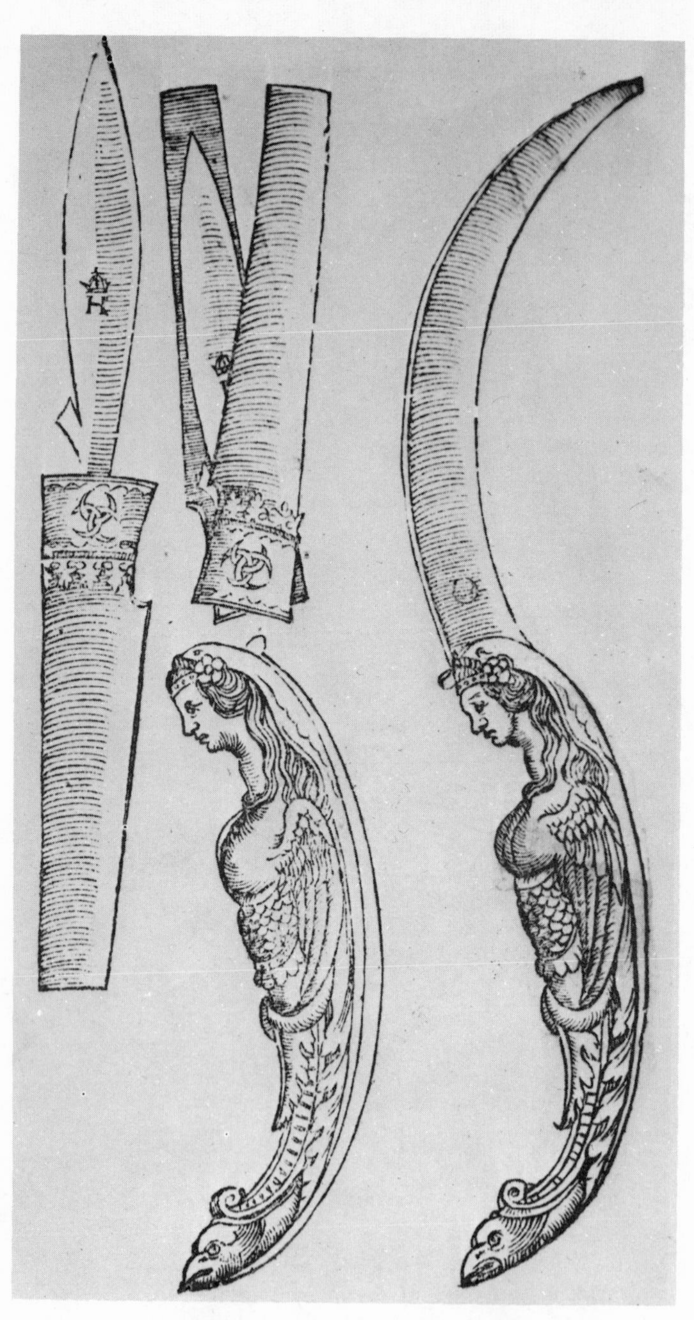

Sixteenth-century surgical knives

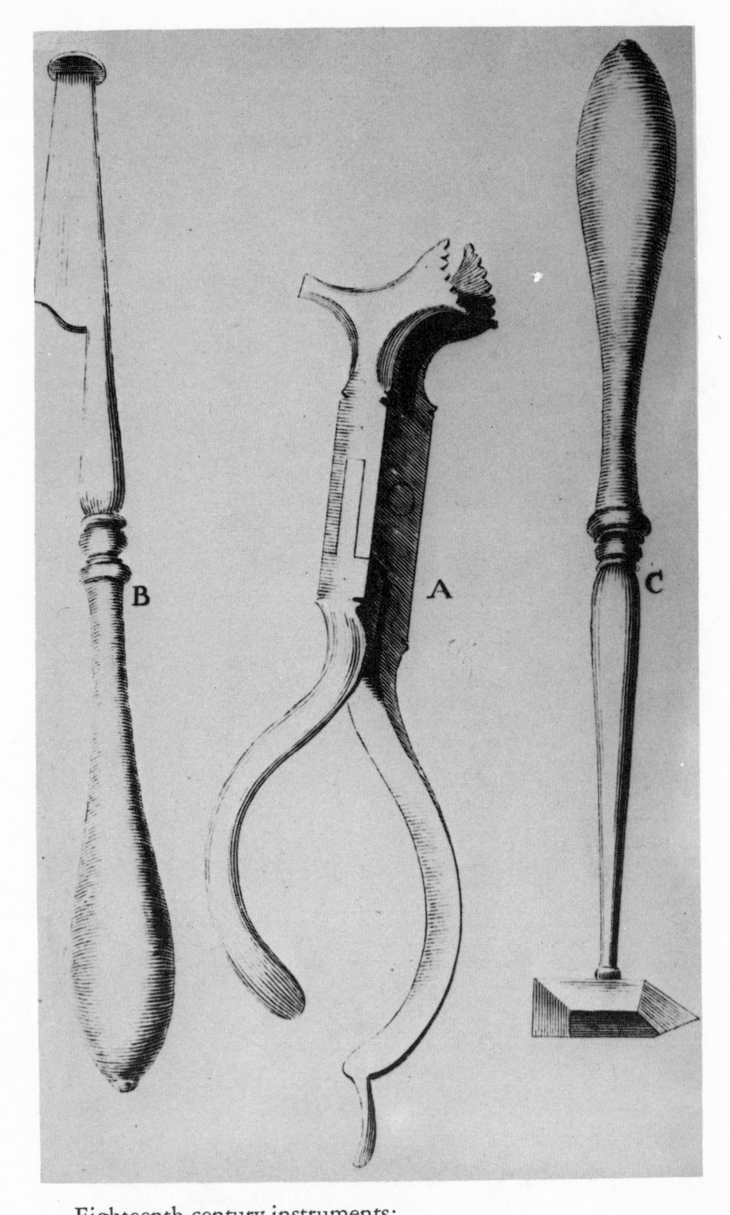

Eighteenth-century instruments:
(A) forceps for removing circular pieces of bone; (B) lenticular with sharp blade for scraping skull and button at end for catching dust; (C) rugine, or raspatory, for scraping bones

Vacuum extractor designed by the nineteenth-century Scottish physician James Young Simpson

Eighteenth-century pewter pap boats

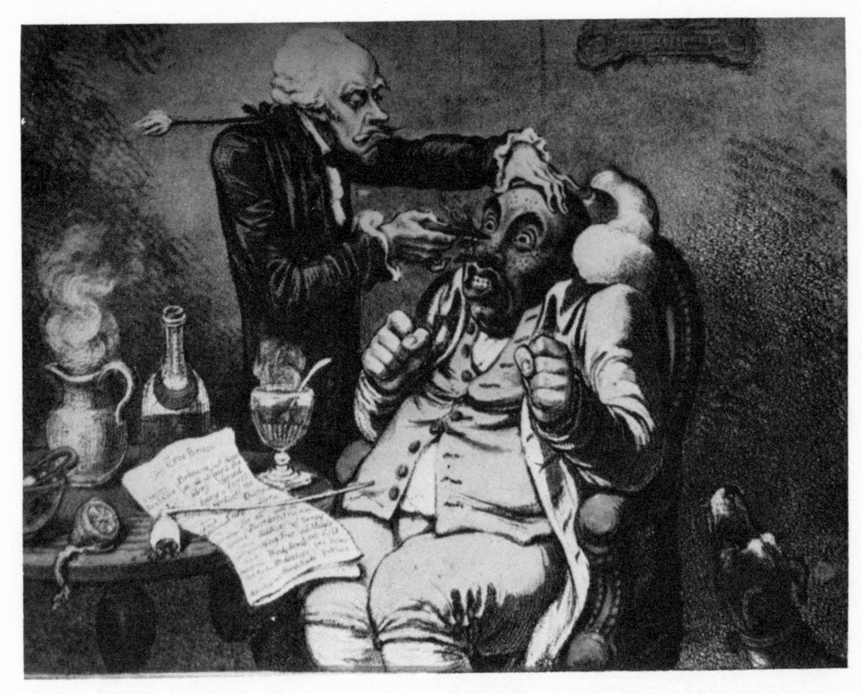

A cartoonist's view of one of the tractors of the eighteenth-century American physician Elisha Perkins, in effective, if not gentle, use

Electric male cystoscope, late eighteenth century

Opposite: Female cystoscope designed by the American surgeon Howard A. Kelly in the 1890s and electric head lamp to be used with it

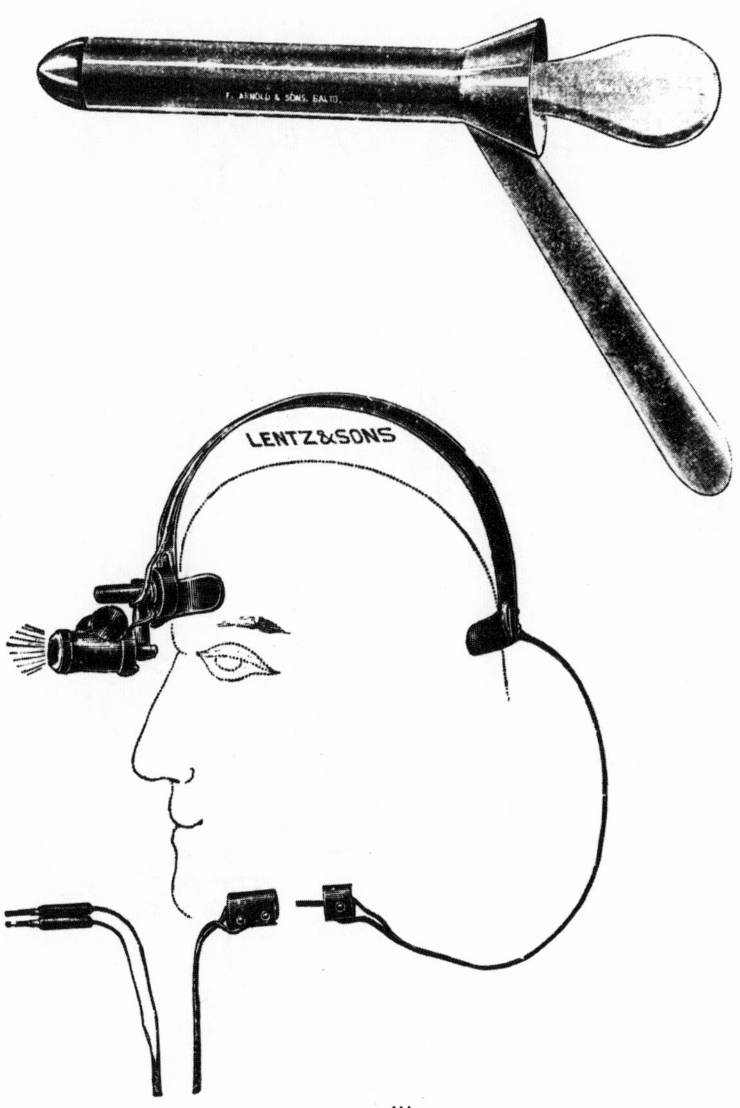
F. ARNOLD & SONS, BALTO.
LENTZ&SONS

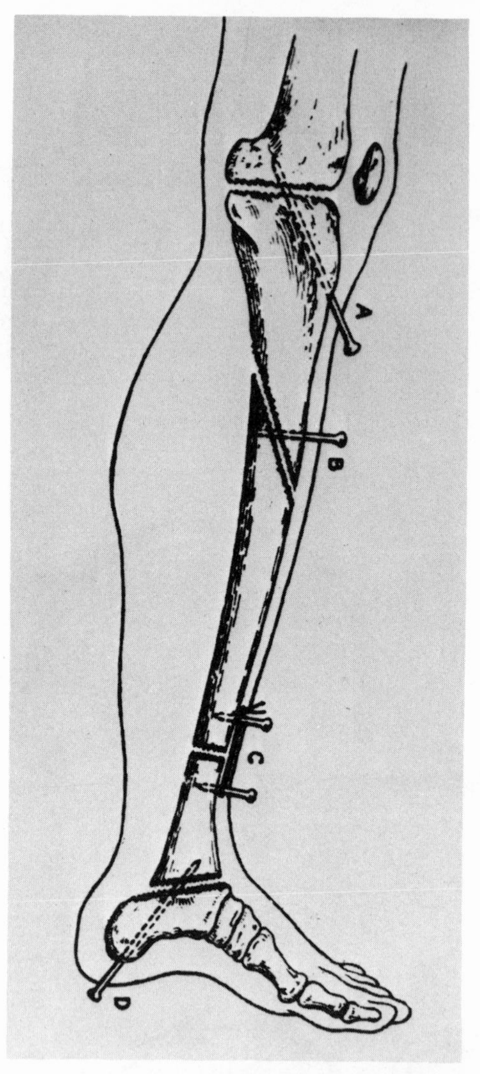

Use of steel pins in late-nineteenth-century surgical practice for: (A) securing bones after excision of knee joint; (B) treatment of an oblique fracture; (C) treatment of ununited fracture of lower end of tibia; (D) securing bones after excision of ankle

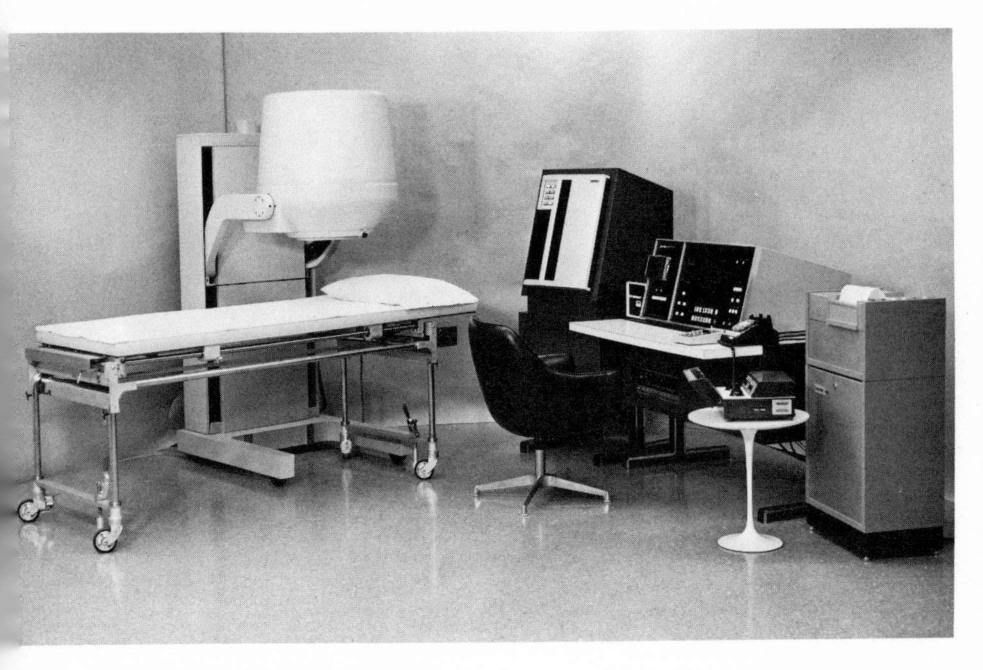

Multicrystal scanning gamma camera for tracing radioisotopes in diagnosis

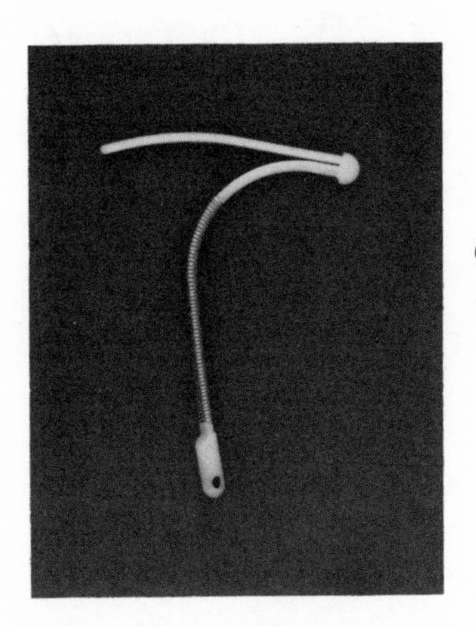

Copper 7 intrauterine contraceptive device, developed in the 1960s

PART ONE

Down the Ages

1

Out of the Crucible

Invite the typical layman to describe a metal and he will likely say that it is solid, relatively hard, shiny (unless it is rusted, in the case of iron, or corroded, in the case of copper), and found in its native state in lumps or nuggets. Asked to name the metals, he will probably start out with aluminum, copper, gold, iron, nickel, platinum, silver, uranium, and zinc (not necessarily in that alphabetical order). He might include mercury but may hesitate to do so because it is a liquid at ordinary temperatures. Pushed, he may add chromium, cobalt, magnesium, molybdenum, titanium, and one or two others. It is unlikely that he will readily name more than a dozen or fifteen metals. Yet of the slightly over one hundred known elements, three-quarters are classified as metals.

Part of the difficulty of recognition lies in the fact that only a handful of metals—gold, silver, copper, platinum, bismuth—are found in a free or native state—that is, *as a metal.* The majority must be extracted from mineral-bearing substances (chemical compounds) called ores.

Few people realize that ordinary table salt is the chloride of a metal—sodium. The soluble white powder celebrated in the play *Arsenic and Old Lace* is a metal. The magnesium flash that accompanied early photography was the ignition of a metal. Anyone who has had to swallow barium in connection with diagnostic intestinal X rays has ingested a metal.

Metals occur with a great range of characteristics, and the basic distinctions between metallic and nonmetallic substances are hardly for the layman. "On a chemical basis," says metallurgist and physicist Bruce A. Rogers, "little distinction can be made. Both metals and nonmetals enter actively into chemical reactions. The difference reveals itself in the physical properties. By common agreement, those elements that possess high electrical conductivity and a lustrous appearance in the solid state are considered to be metals." [1]

Gold, silver, and copper are the metals that most commonly appear in native form. Of these, according to J. Gordon Parr, author of *Man, Metals and Modern Magic,*

> it is likely that gold was the first to be discovered, for it is the most widely distributed. As it is very resistant to corrosion, the bright surface of a gold nugget would attract the attention of an observant primitive man, and no doubt kindle his disgust when it would not chip or break as other stones did. However . . . it made a pretty ornament and might even be temporarily sharpened into a blade.

Because of its rarity and softness, native silver, like native gold, did not lend itself to the making of weapons, but copper was found much more commonly and fabricated much more successfully. Notwithstanding the fact that copper corroded easily, stream erosion or fortuitous scrapings would from time to time reveal the shining metal surface.

> Once discovered, and once some significance had been attached to the discovery, the locality would be searched for more; and . . . native copper might be discovered as very large masses, probably too big to be moved. A lump of copper hacked off would be beaten and hammered until it became a suitable shape. And during this process it became harder. . . . Fortunately for primitive man, copper hardens at ordinary air temperatures; and after hammering it could be sharpened to make a blade or pointed to make a spike . . . and it had the advantage over stone or flint that it was much more easily worked into shape.[2]

In due course man made two vital discoveries: first, that metals could be produced from mineral-bearing rocks; second, that metals could be melted and cast in a mold.

Which discovery came first is an open question, but it seems likely that smelting preceded casting. When primitive man by chance laid his fire on a bed of rocks containing copper, the charcoal produced by the fire reduced the metal. After this phenomenon had occurred a number of times, as it would in a copper-rich region, man was in the copper-producing business.

Civilization had its beginnings around 6000 B.C. in the fertile delta where the Tigris and Euphrates rivers empty into the Persian Gulf. It is reasonable to assume that the art of metallurgy, like that of agriculture, was first practiced there. Progress was halted by the Great Flood (about 4000 B.C.), but when the waters subsided and the surface of the ground dried, peoples from the northern reaches of the rivers again came and settled on the more fertile land. Ornaments and weapons uncovered in the excavations at Ur in Sumer (the southern portion of ancient Babylonia) suggest that the casting of copper, silver, and gold, and the reduction of copper from ore, began around 3500 B.C., with crude, hammered metal articles dating from

2,000 years earlier. A lead statuette in the British Museum indicates that the Egyptians were working with that metal as early as 3400 B.C.

In the early Metal Age man continued to rely on stone implements because copper was generally too soft a metal, but in some areas, where ores contained minerals in addition to copper, smelting produced a harder, stronger metal that was not only easier to toughen but was better to cast. If the additional mineral was cassiterite (tin dioxide), the resultant alloy was similar to what would later be known as bronze. In time man realized that the harder metal could be produced by smelting together copper ores from one locality and tin ores from another. The art of bronze making was learned by the citizens of Hissarlik (on the site of Troy) about 2000 B.C. But tin ores did not occur in nature as universally as copper. This led to two conditions: a delay in the opening of the Bronze Age in certain areas (in Egypt, for example, development of a bronze culture was relatively slow because the Egyptians obtained their copper from the Sinai Peninsula, where tin was not present); trading based on a need for tin.

There is evidence that iron was produced from its ores as early as 4000 B.C. but what might accurately be described as the Iron Age did not have its beginning until about 1400 B.C. A dagger blade dating from 1350 B.C. is probably the oldest surviving iron object created by man. Found in the tomb of Tutankhamen, it was so placed as to suggest that it was a highly treasured possession. A letter written to Pharaoh Rameses II shortly after 1300 B.C. by the king of the Hittites contains the first recorded reference to man-made iron. Steel was first produced on a substantial basis in India during the five centuries preceding the Christian Era.

Classical writers offer the earliest accounts of the production of mercury from cinnabar (mercuric sulfide), but because it occurred in liquid form, they did not regard it as a metal. By the eighth century A.D. mercury had found its place among the metals, but arsenic and antimony, produced from their respective sulfides by the Arabian scholar and alchemist Abu-Musa-Jabor-ibn-Haiyan, or Geber (fl. 721–766), were barred for a pragmatic reason. By then there were seven known metals and each was associated with a heavenly body—gold with the sun, silver with the moon, iron with Mars, mercury (or quicksilver) with Mercury, lead with Saturn, tin with Jupiter, and copper with Venus. Apart from the fact that there was as yet no basis for establishing the characteristics of a metal, there just were not enough known planets with which arsenic, antimony, and other aspirants might have been associated.

It would be a mistake to think that awareness of metals and their potential spread evenly through a developing world. As J. Gordon Parr has pointed out, the Incas were found by the Spanish in the sixteenth century "to have almost no metallurgical knowledge; yet they used native metals very largely. Incredible as it may seem today, many of their axes and knives were of gold, which they valued much as we value iron—for its utility, not because it is scarce." [3] Likewise, the colonists settling in North America found that the natives had neither a desire nor the ability to take advantage of the locality's mineral wealth. "They used native metals, copper and gold; but they neither dug ores nor smelted them. It is almost inconceivable that only 200 years ago so much of the world was so happily uncivilized." [4]

If man was a slow starter he has certainly caught up. Today we recognize so many metals in so many classifications that groups sometimes overlap. The groups include the common metals (those produced in tremendous quantities) and the precious metals; the light metals essential to the transportaton industry; the noble metals (those resistant to high temperature oxidation) and the base metals (those that are chemically reactive); the soft, low-melting, rapidly oxidizing alkali metals, the alkaline earth metals employed principally in compound form, and the rare-earth metals; the semimetals (so-named because of lesser electrical conductivity); and the refractory metals with high melting points essential to the construction of jet engines.

In 1958 Parr could conclude that "aluminum and magnesium and their alloys are in everyday use; the varieties of brass and bronze have become so numerous as to include almost every possible alloying element; zirconium, titanium, uranium are no longer strange names." [5]

2

What Makes a Metal Precious?

While our Stone Age ancestors put copper to a degree of practical use, the use of gold and silver was generally limited to ornamentation. But there is no reason to suppose that primitive man regarded these metals as precious. The ornaments were worn because they were pretty. Times change, however, and what was once taken for granted can acquire an unsuspected value.

> The so-called precious metals include the coinage metals, silver and gold, together with platinum, palladium, and iridium which are used for jewelry.[1]

The first reliable coinage system was established at Lydia in Asia Minor, bordering on the Aegean Sea, around 700 B.C. A standard alloy of gold and silver (electrum) was stamped into coins which became an important element of exchange in the trading (largely in tin) that had developed in the region.

In 490 B.C., notwithstanding the fact that the Athenians had defeated the all-conquering Persian army of Darius I (558?–486 B.C.) at Marathon, the Persian fleet remained a serious menace. Themistocles (527?–460? B.C.), an up-and-coming Athenian statesman and apostle of sea power, proposed the building of a fleet second to none, an undertaking calling for unlimited financial resources. An outcome was the first use of silver as "money" on an extensive scale. "Fortunately, it was at this time that a rich vein of silver was struck in the mines of Laurium, near Athens. The ore was mined, the silver extracted and sold, and ships built." [2]

The Laurium silver occurred with lead. The method used by the Athenians to separate the silver from the lead was known as "cupellation." The alloy was heated in a cupel (a shallow clay basin) and air was blown over the surface. Since lead oxidizes sooner than silver, the lead oxide could be skimmed from the surface of the melt, leaving pure metallic silver. (If there was gold in the original alloy, in addition to silver and lead, cupellation produced the gold-silver alloy electrum. It is uncertain when gold and silver were first separated but a method seems to have been devised around 300 B.C.)

The output of the Laurium mines allowed Athens to establish a uniform silver currency and to assume a dominant role in trading with the rest of Greece and in Asia Minor. Many Greek cities produced coins carrying emblems of a medical nature. As early as 360 B.C. coins minted at Epidaurus and Cos bore the head of Aesculapius, the god of medicine. This practice was continued in Roman times, when coins sometimes depicted Salus, the Roman name for Hygieia, daughter of Aesculapius.

Roman coins were made of brass. The discovery of brass was as accidental as the discovery of bronze. Brass is an alloy of copper and zinc. When zinc ores are smelted alone, they produce zinc vapor; when they are smelted with copper, the copper absorbs the zinc vapor and the end product is brass. Chance seems to have brought the ores together, and there is little evidence that brass was known prior to the Roman Empire. The Bactrians, who dwelt in the northeast section of what was to become Afghanistan and who were contemporaries of the Romans, made copper-nickel coins, with the nickel content running to 20 percent and more.

Modern currencies have employed gold, silver, nickel, and copper. At the beginning of the present century, gold and silver coins represented their face value in the metal involved, but the economic ills of the past fifty years have seen the withdrawal of gold coins and the debasement of so-called silver coins.

An unusual coinage was introduced at the leper colony on Culion Island in the Philippines:

> To avoid the use of ordinary legal tender by the patients, the government in 1913 provided for a special coinage backed by an equivalent amount set aside in the treasury. The first issue, of six denominations, . . . was of aluminum, as also was a further partial issue made in 1920. That metal was found disadvantageous because of corrosion by antiseptics; when mercuric bichloride was used the coins simply disappeared. The more recent (partial) issues, in 1922, 1925, 1927 and 1930, were all nickel. All payments made to the inmates were—and still are—made in this money. At first many of the patients refused to accept it and law suit and even physical violence was threatened; not a little, in contempt, was thrown into the sea. In due course this attitude changed.[3]

When King Hiero II of Syracuse in Sicily, a Greek colony, suspected that his goldsmiths had alloyed with silver the gold he had furnished them to make a crown, he required Archimedes (c. 287–212 B.C.), a mathematician and inventor of Syracuse, to determine whether he was right. "We shall never know whether an inspired guess or an involved process of logical thought led Archimedes to realize while bathing that a body displaces water according to its volume and not according to its weight. So for equal weights, a light metal would displace more water than a heavy one; the gold-silver alloy of the crown should displace more water than pure gold of the same weight." [4]

The king's suspicions turned out to be justified, but, more important, Archimedes had taken a first step toward an understanding of metal structure.

It was lust for gold that led to the Spanish conquests in the Americas. Columbus's objective was to find a direct route to China and India, and he believed he had attained this objective when, in what was actually the Bahamas, he found natives adorned with gold ornaments, who told him more gold was to be had toward the south. Putting in at Cuba and Haiti, he seized gold culled from the stream beds by local natives. Subsequent to Columbus's death in 1506, a Spanish expedition landed in Mexico and was met by ambassadors of Montezuma (1480?–1520), whose request that the expedition leave was backed by extravagant gifts. The effect was the reverse of what Montezuma had hoped. Allured by the pricelessness of the intended bribe, the Spaniards invaded Mexico in search of additional treasures. A further promise of gold did not deter them. Finally, after a hard-fought battle, the Spanish under Hernando Cortez (1485–1547) conquered Mexico in August 1521.

In an immediate sense the brutality and massacres involved were in vain. The gold resources of the country failed to come up to expectation. Cortez and his followers had mistaken the treasures accumulated by Montezuma and his predecessors over many years for current production. However, within twenty years rich Mexican deposits of gold and silver were found and worked—with a portion of the output earmarked for the king of Spain.

Peru fell to the Spanish in 1531, and within a few years the west coast of South America was in Spanish hands. Gold and silver were mined in productive areas by native slave laborers who were subjected to treatment that would have seemed brutal to Athenians and other early peoples who employed slaves in their mines. The only solace offered the Peruvian Indians was a local herb called coca (cocaine) which "enabled men to go without food, water, and sleep, to keep warm while others froze, and to remain fresh after hours of work." It was not long before the Spanish conquerors were feeding coca to the slave miners as a means of subduing them and keeping them happy. After all, an Indian on coca could be counted on to work harder and longer.[5]

A 1606 charter granted by James I of England to colonists headed for America stipulated that the Crown should receive one-fifth of the precious metals and one-fifteenth of the copper found, a stipulation that seems to have proved more optimistic than realistic.

Gold, platinum, silver, copper, and a few other metals are still used in fashioning jewelry, from the costly to the inexpensive. Silver as a medium of investment or exchange has become greatly depreciated in value. Gold with its two-price structure has become a pingpong ball of international speculation.

Persian-born ibn Sina, or Avicenna (980–1037), known to his contemporaries as the Prince of Physicians, was an eminently successful practitioner at the court at Baghdad. His professional standing lends credence to a story that he gave his elite patients pills coated with gold and silver, not only to make them more palatable and more attractive but also—according to George Edward Trease of the University of Nottingham, England—to increase their therapeutic efficiency. (Trease adds that the practice of silvering pills was quite common in his youth.[6]) Avicenna is also credited with employing gold in its metallic state as a purifier of the blood.

Such legends illustrate two uses of precious metals in medicine: the use in medicine of metals simply because they are classified as precious and for no sound therapeutic reasons and the use of precious metals for their accredited therapeutic value. In addition, some nonprecious metals have become *precious in medicine* because of their therapeutic effectiveness.

The use of metals in medicine is not, of course, restricted to therapy. They are employed in the manufacture of instruments and prostheses, in implants, and in diagnosis.

3

The Early History of Metals in Medicine

The Old Testament makes no mention of the medicinal use of metals. It is true that no less an authority than the Bavarian chemist Georg Ernst Stahl (1660–1734), the apostle of animism, supposed that Moses had therapeutic intentions when "he took the calf which they had made, and burnt *it* in the fire, and ground *it* to powder, and strawed *it* upon the water, and made the children of Israel drink *of it,*" [1] but, comments John Henry Pepper in his *Play Book of Metals,*

> Not the least intimation is here given of the gold having been dissolved, chemically speaking, in water. After the form of the calf had been destroyed by melting in the fire, it was stamped, and ground, or, as the Arabic, and Syriac versions have it, filed, into a fine dust, and thrown into the river, of which the children

> of Israel would drink. Part of the finely-powdered gold would remain, notwithstanding its greater specific gravity, suspended for a time on the surface of the river; in which condition the gold might be swallowed, distastefully indeed, but harmlessly, together with the water, in the manner described. If actually the Israelites had drank gold in a state of solution they might have thus imbibed a rank poison.[2]

Early medical treatment was based largely on superstition; the small balance was empirical. "When did a patient call a physician and when a priest-magician," asks medical historian Henry E. Sigerist. "This is difficult to tell. The complaint, the illness, may have been the determining factor. When a man suddenly raved or said strange words in a delirious condition an incantation seemed the most appropriate treatment. Or a patient in certain cases might have consulted both, first the physician then the priest or vice versa. . . . It may also be that economic considerations played a certain part in the choice of healer. Remedies often contained rare drugs that were costly, while an incantation could be paid for with a modest offering to the temple." [3]

A favorite remedy of the priest-magician was the amulet charged with the magic power of incantation. Amulets for the poor might be "made of fishbones, a linen bag, or similar cheap materials and . . . for the rich . . . of beads of gold and precious stones." (Sigerist notes this use of beads of gold in both Mesopotamia and Egypt.) The latter "were amulets for the children of rich parents. The choice of stones and metals was by no means arbitrary, and the selection was not made primarily because the materials were precious, although this was a factor since it implied the idea of sacrifice. The choice was rather determined by the fact that in Egypt as in all ancient civilizations minerals, metals, plants, every object, animate and inanimate, were attributed certain magic virtues. The correct choice and combination of materials were therefore extremely important." [4]

There were no pharmacists in ancient Egypt and the number of metal remedies compounded by the physicians was limited. Antimony sulfide, copper acetate, copper sulfate, copper carbonate, and sodium carbonate were applied in the treatment of the eyes. "Among the drugs named in the [Ebers] papyrus . . . are certain substances, evidently metals by the suffixes, but they have not been exactly identified. Neither gold, silver, nor tin is included. One is supposed to be sulphur, another, electrum (a combination of gold and silver), and another alluded to as 'excrement divine,' remains mysterious. Iron, lead, magnesia, lime, soda, nitre and vermilion are among the mineral products which were then used in medicine." [5]

This papyrus, discovered by the German Egyptologist Georg Ebers in 1872, is believed to date from 1552 B.C., and offers an unusually complete account of medicine and pharmacy in ancient Egypt. The Babylonians and Assyrians appear to have been ahead of the Egyptians in employing metals in medicine. "Among the chemical elements and compounds encountered in Assyrian prescriptions, we find white and black sulphur, sulphate of iron, arsenic, yellow sulphide of arsenic, arsenic trisulphide, black saltpeter, antimony, iron oxide, magnetic iron ore, sulphide of iron, pyrites, copper dust, verdigris, mercury, alum, bitumen, naphtha, calcinated lime and a variety of not identified stones." [6]

According to an unidentified seventeenth-century writer:

> The antients put leaf-gold into many compositions; but I know not for what end but to feed the eye; for its substance is too solid, and compact to be dissolved by our heat, and brought into act; nor is it available that some make the vertues, or spirits of Gold sympathising to those of the heart, and therefore give leaf-gold;

> for, by the same facility it may destroy the heart. And it may be apply'd outwardly in greater quantity, and with more profit, with little, or no inconvenience.[7]

There is evidence that gold was used curatively by the Chinese two thousand years before Christ. In any event, the writings of the Taoist philosopher Pao Pu Tzŭ (253–333?) make it clear that metals played an important role in the search of the Chinese for immortality—a condition which must at least to some degree be equated with good health and longevity.

According to Pao Pu Tzŭ, the highest ranking "medicine of the immortals" was cinnabar (red mercuric sulfide). "Take three pounds of genuine cinnabar, and one pound of white honey," he wrote in his *Nei P'ien*. "Mix them. Dry the mixture in the sun. Then roast it over a fire until it can be shaped into pills. Take ten pills the size of a hemp seed every morning. Inside of a year, white hair will turn black, decayed teeth will grow again, and the body will become sleek and glistening. If an old man takes this medicine for a long period of time, he will develop into a young man. The one who takes it constantly will enjoy eternal life and will not die." [8] Cinnabar was followed in descending order of effectiveness by gold and then silver. "Cinnabar and gold were considered as medicines *par excellence,* and were frequently used in combination with excellent results in the attainment of immortality. Preparations in which silver was the chief ingredient had only a limited degree of efficacy in producing immortality. . . . Cinnabar was also a favorite ingredient in life-prolonging concoctions, by virtue of its producing mercury, the 'living metal,' when subjected to heat." [9]

The British pharmacologist-historian C. J. S. Thompson found it "somewhat curious that while the alchemists of the West were always in doubt as to what constituted the true Philosopher's Stone, the Chinese seemingly had no doubt as to its identity. Cinnabar was regarded by the early alchemists and philosophers of that nation as the wonderful body which was supposed to have the mysterious power of converting other metals into gold, and when used as a medicine would prolong life for an indefinite period." [10]

The Chinese prized mercury as an elixir, while Indian physicians of the Brahman period (800 B.C.–A.D. 1000) employed it for skin diseases, including smallpox, and later for syphilis. Other metals in the Hindu materia medica included gold, silver, copper, tin, lead, zinc, arsenic, iron, sodium (in the form of borax), and aluminum (in the form of alum).

The Greek physician Hippocrates (460–377 B.C.) drew most of his drugs from the vegetable and animal kingdoms, but he made use of alum and some copper and lead derivatives. Thompson, while not directly attributing its use to Hippocrates, includes cinnabar, "which seems to have been known from a very remote period," among the "chemical bodies and drugs known both to the Greeks and the Romans." [11]

The Roman historian Pliny the Elder (23–79), whose untimely death resulted from "his eagerness to observe" the eruption of Mount Vesuvius that buried Pompeii (he was "suffocated by the gases" when he "approached too near"), [12] devoted thirteen books of his *Natural History* to medicine, a work that was "interesting for its many curious facts about plants and drugs, its sidelights on Roman medicine, and its author's many slaps at physicians." [13] Most of the minerals used in Roman medicine—iron, lead, nitrum, salt, gold, tin, silver, realgar (arsenic sulfide), copper, and misy (a combination of the sulfates of copper and iron)—were known to Pliny.[14]

Pliny reported that pure gold applied externally removes styes on the eyelids. A popular adoption of this was in vogue as late as 1907 "as shown by the common practice of wetting a gold ring with fasting saliva, and then gently rubbing the ring thus moistened, (and probably superficially dissolved by the saliva), along the outside of the eyelid over the stye." [15] Marcus Terentius Varro (116–27 B.C.), a celebrated encyclopedist who wrote on a variety of subjects, including medicine, considered gold a cure for warts.

Galen (130–c.200), the greatest Greek physician after Hippocrates, established a practice in Rome in the year 164 but, two years later, at the height of his success, abandoned it in favor of travel, study, and teaching. He "visited the copper mines at Cyprus to obtain diphryges (probably a by-product of zinc oxide), misy, sory, or chalcanthos (different names for crystallized iron sulphate) which was used as an astringent, cadmia (zinc oxide?), copper ore, and pompholyx ('flowers of zinc')." [16]

> When washed, pompholyx [Galen said] is an excellent remedy for drying out in an unirritating way, and is especially useful in cancerous and other malignant ulcers. On account of its non-irritating quality it is used in collyria for inflammations of the eye, like phlyctenulae and corneal ulcers.[17]

Metal Surgical Instruments

"The earliest surgical instrument was, in all probability, . . . some fragment, unusually sharpened as to edge and point by accidental flaking, as in the obsidian knives of Peru. By means of these sharpened flints or of fishes' teeth, blood was let, abscesses emptied, tissues scarified, skulls trephined. . . . In the Bronze Age surgical saws and files were plentiful everywhere, from Egypt to Central Europe. . . . Some time later, as, for instance, among the Gallo-Roman finds in France, we can trace the evolution of the jointed or articulated surgical instruments, like scissors, in which cutting was done by indirect action. With improved metal instruments, such cosmetic operations as tattooing, infibulation, boring holes for ear-rings and nose-rings, or the Mica operation (external urethrotomy), as well as amputation and lithotomy, could be essayed. The ancient

Hindus performed almost every major operation except ligation of the arteries."[18]

The Hippocratic writings of the fifth century B.C. contain the earliest descriptions of surgical instruments made of metal. There was the trephine, a saw with a circular motion used in treating skull injuries, the bone drill, employed when the area of disease or of a wound was larger than could be attacked with a trephine, and the cranioclast, the use of which was thus described by Hippocrates: "Opening the head with a scalpel, break it up with the cranioclast in such a way as not to splinter it into fragments, and remove the bone with a bone forceps."[19]

There were various probes or sounds including an ointment spatula, a uterine sound for correcting malpositions of the uterus and dilating and applying medications to the interior of the cervix, and a graduated set of dilators made of wood, tin, or lead.

Uvula forceps were employed to crush the uvula prior to amputation—and later, according to Paul of Ægina (625–690), to clamp hemorrhoids to facilitate their removal with a scalpel. Tooth forceps existed, but the ancients, because of a number of associated deaths, regarded tooth extraction as an operation to be avoided. (According to Caelius Aurelianus, an African neurologist of the fifth century A.D., a tooth forceps made of lead was hung in the temple of Apollo at Delos to remind operators to exert little force in tooth extraction.) Needles were used for stitching and securing bandages and splints. On the subject of complicated fractures, Hippocrates described this procedure:

> In those cases of fracture in which the bones protrude and cannot be restored to their place, the following mode of reduction may be practised: pieces of steel are to be prepared like the levers which the cutters of stone make use of, one being rather broader and the other narrower, and there should be at least three, or even more, so that you may use those that suit best, and then along with extension we must use these as levers, applying the

> under surface of the piece of iron to the under fragments of bone, and the upper surface to the upper bone, and in a word we must operate powerfully with the lever as we would do upon a stone or log. The pieces of steel should be as strong as possible so that they may not bend.[20]

After the Hippocratic contribution, the well of medical writings ran dry for four hundred years, until Aurelius (Aulus) Cornelius Celsus (25 B.C.–A.D. 50) produced his eight books on medicine. Celsus, a Roman aristocrat and almost certainly not a physician, has been hailed as the foremost medical writer of his day.

It was inevitable that the discovery of metal should lead to the development of surgical instruments and it was equally inevitable that the instruments of Celsus' day should be an improvement over the instruments described by Hippocrates. The art of surgery could not, of course, reach full maturity until the introduction of anesthesia, antisepsis, radiography, and safeguards against shock, but if the array of surgical instruments found in the excavations of Roman remains at Pompeii (including the House of the Surgeon, which yielded a fine collection now exhibited at the Museo Nazionale in Naples) and Herculaneum—and in such outposts of the Roman Empire as Paris, Rheims, Fonviel, Namur, Baden, and Cologne—is any criterion, surgery (especially superficial surgery) took a great step forward at the beginning of the Christian Era. "Surgery was highly refined as long as the patient had courage and the doctor had good tools and experience, and the head and the abdomen were not involved." [21] Celsus emphasized the importance of surgery in the training of the conscientious physician, a view that would subsequently be shared by Galen.

Roman surgical instruments may be divided into seven general categories—knives (or cutting instruments), probes, forceps, bleeding cups, instruments for cauterizing, bone and tooth instruments, and bladder and gynecological instruments.

"The surgical knife had, as a rule, the blade of steel and the handle of bronze. We find specimens all of steel or all of bronze but these are exceptional forms; and hence it happens that many more handles than blades have been preserved to us, as usually the blade has oxidized away leaving no trace of its shape." [22] (In fact, most of the surgical instruments that have survived are made of bronze.) The blades of the knives were straight or curved, often with two cutting edges. Two-edged curved blades were always sharp-pointed. Otherwise knives might be sharp- or blunt-pointed. In shape, the scalpel did not change much from Hippocrates' time down to the twentieth century.[23]

Scissors were developed early and Celsus, among others, refers to cutting the hair as a therapeutic measure. There are, however, few references to the use of shears for severing tissue (possibly because smooth enough edges to the blades could not be achieved). But Celsus suggests an exception:

> There have been others who cut away the omentum with scissors, which is unnecessary if the portion is small: and if very great it may occasion a profuse haemorrhage, since the omentum is connected with some of even the largest veins. But this objection cannot be applied in cases where, the belly being cut open, the prolapsed omentum is removed with shears, since it may be both gangrenous and unable to be removed in any other way with safety.[24]

Probes were "a very comprehensive class. The original specillum was no doubt a simple sound. . . . But the custom of combining two instruments on one shaft gradually led to the application of . . . the term specillum to denote a large variety of instruments." [25] For example, a spatula, spoon, or hook might be found on the other end of a probe.

The Romans were well aware of the value of information

obtained by exploring the recesses of a lesion with a rod of metal.

> But first it is well to put a probe into the fistula to learn where it goes and how deeply it reaches, also whether it is moist or rather dry as is evident when the probe is withdrawn. Further, if there be bone adjacent, it is possible to learn whether the fistula has entered it or not and how deeply it has caused disease. For if the part is soft which is reached by the end of the probe the disease is still inter-muscular; if the resistance be greater it has reached the bone: if there the probe slips there is as yet no caries. If it does not slip but meets with a uniform resistance there is indeed caries, but it is as yet slight. If what is below is uneven and rough the bone is seriously eroded, and whether there is cartilage below will be known by the situation, and if the disease has reached it will be evident from the resistance.[26]

Uvula and tooth forceps date back to Hippocratic times. Galen describes a polypus forceps for removing nose tumors. It had a forceps at one end and a scoop for scraping away the tumor at the other. The myzon, or tumor forceps, was used whenever it was necessary to exert a pull on a tumor or some other object in order to excise it, or to raise a piece of skin. The jaws of the myzon were finely toothed. Of the pharyngeal forceps Paul of Ægina wrote: "Prickles, fish-bones and other substances are swallowed in eating and stick in different places. Wherefore such as can be seen we are to extract with the special fish-bone forceps." However, says Aetius of Amida (fl. 550), "bones stick near the tonsil or back of the pharynx and can be seen, and if a considerable part projects out of the tonsil it can be removed with an epilation forceps." [27]

The extraction of blood by means of cups dates to remotest times. Celsus describes two kinds of cups, bronze and horn:

> The bronze is open at one end and closed at the other, the horn, open at one end, as in the previous case, has at the other end a

small foramen [opening, perforation, or orifice]. Into the bronze kind burning lint is placed, and then the mouth is fitted on and pressed until it sticks. The horn one is placed empty on the body, and then by that part where the small foramen is, the air is exhausted by the mouth, and the cavity is closed off above with wax, and it adheres in the same way as before. Either may advantageously be constructed not only of these varieties of material but of any other substance. If other things are not to be had a small cup or narrow mouthed jar will answer the purpose. When it has fastened on, if the skin has previously been cut with a scalpel it extracts blood; but if it be entire, air.[28]

Indirectly related to the bleeding cup is the cannula for withdrawing an accumulation of fluid containing serum: "We introduce through the incision in the abdomen and peritoneum, a bronze cannula having a tip like that of a writing pen." [29] A similar tube, made of tin, was used by Hippocrates to extract pus from the pleural cavity.

Cauterizing instruments were common in ancient times. They were nearly always made of iron because bronze became too soft when used as a cautery. Hippocratic writings refer to cauteries as "the irons." There were, of course, exceptions. Theodorus Priscianus (fl. A.D. 380) recommended cauteries of gold or silver for stopping hemorrhaging of the throat. The cautery was used for a great variety of purposes—as a counter-irritant, as a hemostatic, as a bloodless knife, as a means of destroying tumors, just to name a few.

The raspatory or rugine consists of a blade of varying shape fixed at right angles to the shaft, and it is operated by pulling instead of being driven forwards by striking or pushing. Although no ancient raspatory has been preserved to us we are quite familiar with the instrument, as it has been in continuous use through ancient and mediaeval times, and it is in use at the present day. The raspatory is the instrument upon which Hippocrates relies for eradicating fissured and contused bone in injury to the skull.[30]

Raspatories came in different sizes and shapes. A small one was used as a tooth scaler. Other bone instruments included several types of chisel employed in the repair of skull injuries, a hammer for use with the chisels, a meningophylax, or small plate, for inserting under a bone which is being cut in order to protect underlying structures, drills, saws, and trephines, the bone lever of Hippocrates already described, bone forceps, tooth forceps, files, and forceps for extracting imbedded weapons.

Bladder and gynecological instruments included catheters, both for releasing urine and pushing back a stone impacted in the urethra, the lithotomy scoop and forceps for extracting calculi revealed by an incision, rectal and vaginal specula for dilation of the rectum and vagina, a traction hook for removal of the fetus in a difficult labor, and a uterine curette.

Obviously the technical competence of the Roman surgeons improved with the development of tools in new shapes and the discovery of new metals and alloys which were not only less costly but provided sharper cutting edges. The Romans became unusually skillful in the manufacture of small instruments. Celsus, Scribonius Largus (fl. first century A.D.), Galen, and Marcellus Empiricus (fl. c. A.D. 400) all mention the "strigil" as being used to get into small openings. "After having heated the fat of a squirrel in a strigil, insert it into the auditory canal," says Galen.[31]

According to Galen, the best steel for the making of surgical tools came from Noricum, in what was to be Austria and southern Germany. "In the case of the knife or a scalpel, the instrument sought was one not easily blunted, chopped, or bent. In the making of such tools, the Roman founders searched for the finest ores, containing at least 75 per cent iron, and then working with a charcoal fuel, nearly pure carbon, they produced a fine quality of steel in limited amounts as easily as they could iron." [32] Nevertheless, many Roman surgical tools were still made of bronze. This may be attributed to the fact that copper ore was more readily available to the Romans than iron ore. Bronze tools cost less.

The Hindu physician Suśruta (c. A.D. 500) described 127 surgical instruments, mostly made of steel. He gave special attention to the condition in which they should be maintained, indicating that cutting instruments should be sharp enough to split a hair lengthwise.

Other metals sometimes used in the fashioning of surgical instruments were tin, lead, gold, and silver, some uses of which have already been mentioned. Hippocrates refers frequently to uterine sounds made of tin and speaks of tin sounds and eyed probes for rectal work, this metal being preferred because of its flexibility. Hippocrates used sounds and tubes of lead for intra-uterine medication, and Celsus and Paul refer to leaden tubes for insertion in the rectum and vagina to prevent contractions and adhesions of scar tissue following surgery. Hippocrates used gold wire to bind teeth together in jaw fracture cases. The great Moslem physician Avenzoar, writing in the twelfth century, speaks of a golden probe for applying salve to the eye and for separating adhesions of the eye to the lid, and Avicenna evacuated smallpox pustules with a golden probe. Johannes Mesuë (777–837), the Christian director of the Nestorian hospital at Djondiapour in Persia, recommended a heated scalpel of gold to excise a tonsil. Silver was employed with less frequency, but Hippocrates describes a uterine syringe with a silver tube and Albucasis (1161–1231) mentions silver catheters. Drugs were stored in copper bowls, in tin, bronze and silver boxes, and in leaden jars.

4

The Role of the Alchemists

"Alchemy," says the English chemist H. Stanley Redgrove, author of several books relating science to philosophy and mysticism, "is generally understood to have been that art whose end was the transmutation of the so-called base metals into gold by means of an ill-defined something called the Philosopher's Stone; but even from a purely physical standpoint, this is a somewhat superficial view. Alchemy was both a philosophy and an experimental science, and the transmutation of the metals was its end only in that this would give the final proof of the alchemistic hypotheses; in other words, Alchemy, considered from the physical standpoint, was the attempt to demonstrate experimentally on the material plane the validity of a certain philosophical view of the Cosmos. . . . Unfortunately, however, not many alchemists came up to this ideal; and for the majority of them, Alchemy did mean merely the possibility of making gold cheaply and gaining untold wealth." [1] Or, as an earlier writer put it, "Alchemy aimed at giving experimental proof of a certain theory of the whole system of nature, including humanity. The practical culmination of the alchemical tests presented a threefold aspect: the alchemists sought the stone of wisdom, for by gaining that they gained the control of wealth; they sought the universal panacea, for that would give them the power of enjoying wealth and life; they sought the soul of the world, for thereby they could hold communion with spiritual existences, and enjoy the fruition of spiritual life. The object of their search was to satisfy their material needs, their intellectual capacities, and their spiritual yearnings. The alchemists of the nobler sort always made the first of these objects subsidiary to the other two." [2]

But however noble or ignoble the aims of the alchemists "it was chiefly alchemy that was responsible for the steady improvement of the technical apparatus which was to become indispensable to the preparation of medicines. . . . For . . . in the course of their experiments [the alchemists] discovered and amassed such a treasure-house of chemical data that alchemy can deservedly be called the forerunner of chemistry." [3]

When Did Alchemy Originate?

The first name associated with the history of alchemy is that of Hermes Trismegistus. Alchemists who came later believed him to be an Egyptian contemporary of Moses, but it is now generally accepted that no such historical person existed and that several works attributed to him were written by one or more later authors.

There is abundant evidence that the art which later acquired the name of alchemy made its first Occidental appearance in Greek-speaking Egypt a little before or after 100 A.D. For such a

> date the most important and decisive testimony is that of Pliny the Elder (23–79 A.D.), whose *Natural History* covers every existing field of investigation, shows a special interest in metallurgy, medicine and magic, and forms in fact the finest single source for the historical background of alchemy, but does not reflect a single idea that can properly be called alchemical. His book reveals a medley of aspirations, beliefs and superstitions, amongst which the idea of the transmutation of metals was apparently ready to burst into bloom, but the bud had not yet opened.[4]

The name Hermes Trismegistus first appeared in a third-century A.D. papyrus from the Greek port city of Hermopolis.[5] As to the origin of the name, it involved identification of the Egyptian god Thoth with the Greek god Hermes. Both were connected with the dead, Thoth being the "advocate of the dead" on the Day of Judgment, "as weigher of the heart, as recorder of the verdict," and Hermes being the messenger of the gods, who escorted the souls of the dead to the underworld.[6] (As early as the fifth century B.C. Herodotus spoke of the city and temple of Hermes rather than the city and temple of Thoth; six centuries later Plutarch regularly referred to Thoth as Hermes.) There has been considerable speculation as to the significance of "Trismegistus," which is derived from a Greek word meaning "three times greatest." It may be an equivalent of an Egyptian expression "very great-great," which was sometimes applied to Hermes.

It was Geber (fl. eighth century A.D.), however, who came to be regarded as the founder of alchemy. While there can be "no dispute that with the name *Geber* was propagated the memory of a personality with which the chemical knowledge of the times was bound up," [7] there were unquestionably alchemists of the Greco-Egyptian-Arabian world actively engaged in their art before Geber, ranging from the third-century Zosimus through Synesius of Ptolemais (fl. 410) to historian Olympiodorus of Thebes (fl. 395–423).

Of a number of treatises on alchemy written by Zosimus, only fragments have survived. These fragments, however, "contain descriptions of apparatus, of furnaces, studies of minerals, of alloys, of glass making, of mineral waters, and much that is mystical, besides a good deal referring to the transmutation of metals." [8]

Alchemy was also practiced in China. Pao Pu Tzŭ wrote in the third and fourth centuries A.D. that, while the indefinite prolongation of life may have been the primary objective of alchemy, its second phase was concerned with making life more comfortable. "While its aim in general was the transmutation of base metals into those of high value, its aim in particular was the manufacture of gold. For gold was then, as now, a synonym for riches, and its possession a sufficient guarantee for a life of comfort and ease." [9]

The Early Alchemists

The idea that all metals were (essentially) the same thing in varying degrees of purity had been advanced before the time of Geber.

> The alchemists thought that gold was the purest and noblest of all metals and that it "ripened in the earth." The dream of the alchemist was to hasten this supposititious natural process. Amalgamation had long puzzled them, and the idea gradually got abroad that mercury was very closely related to gold and, in fact, only lacked the addition of a mysterious "something" to give it the proper color and solidity. This missing "something" was supposed to be related to sulphur, probably because of its color. The search for the proper tinctorial substance, or tincture of metal, or "philosopher's stone," as it was more commonly called, occupied the energies of many men. The Arabic name for the "philosopher's stone" (*al-kibrīt al-ahmar*) literally means the red sulphur.[10]

But hand in hand with the idea of the transmutation of metals always went the notion of a cure-all, a fountain of youth, an elixir of life. This elixir was supposed to be in the nature of drinkable gold (*aurum potabile*). "The search for potable gold led to the discovery of aqua regia [nitrohydrochloric acid water] and the strong acids by Geber and Rhazes [850–923], and the quest of the elixir was the starting point of chemical pharmaceutics." [11]

After Rhazes and Avicenna (in the eleventh century), the teachings of the Arabian alchemists moved into the western world. Writings on alchemy have been attributed to Albertus Magnus (Albert von Bollstädt, 1193?–1280) and his pupil Thomas Aquinas (1225–1274), both celebrated members of the Dominican order, but the authenticity of their authorship is questionable. Consequently, the title of "most illustrious of the mediaeval alchemists" [12] has been conferred on English-born Roger Bacon (1214–1294), who studied in Paris, very likely under Albertus Magnus. It has been said of Roger Bacon,

> He is to be regarded as the intellectual originator of experimental research, if the departure in this direction is to be coupled with one name—a direction which, followed more and more as time went on, gave to the science [of chemistry] its own peculiar stamp, and ensured its steady development.[13]

Bacon was a systematic thinker, capable of gathering extensive data in such a way that he was able to explain apparent inconsistencies. Thoroughly familiar with alchemy and its literature, he believed in the possibility of the transmutation of metals and also in the philosopher's stone. In fact, he viewed medicine as a means of prolonging life through alchemy. "*Alchimy,*" he wrote, "is a Science, teaching how to transforme any kind of mettall into another: and that by a proper medicine, as it appeareth by many Philosophers Bookes. *Alchimy* therefore is a science teaching how to make and compound a certaine medicine, which is called *Elixir,* the which when it is cast upon mettals or imperfect bodies doth fully perfect them in the verie proiection." He subscribed to an earlier view that all metals are composed of two elementary principles—sulfur and mercury—combined in different proportions and achieving different degrees of purity. He maintained that "the naturall principles in the mynes, are *Argent-vive,* and *Sulphur.* All mettals and minerals, whereof there be sundrie and divers kinds, are begotten of these two: but I must tel you, that nature alwaies intendeth and striveth to the perfection of Gold: but many accidents comming between, change the mettalls. . . . For according to the puritie and impuritie of the two aforesaide principles, *Argent-vive,* and *Sulphur,* pure, and impure mettals are ingẽdred." [14]

The Catalan Arnold of Villanova (1235–1311), the "genius of the West at that time . . . who made several important discoveries in chemistry" and whose "researches were particularly directed to the relation of chemistry to medicine," [15] was a disciple of the Arabian chemists and of the earliest European writers on alchemy. With doctorates in theology, law, philosophy, and medicine, he was recognized as a skillful physician and is credited with introducing *aurum potabile* into the pharmacopeia.

If the stories about him are to be believed, Raymond Lully (1235–1315), born on the island of Majorca, had a checkered career. Starting out as a libertine, he was so shaken by the sight of the cancerous breast of a would-be mistress that he decided to become a missionary and convert to Christianity the (Moslem) heathens of Africa. (He was ultimately stoned to death by the inhabitants of Bugiah in Algeria.) While his knowledge of medicine seems to have been superficial, a number of books on alchemy have been credited to him. In his *Clavicula* [*A Little Key*], he attacked the alchemy of his day, making it clear that he believed in multiplication of the noble metals rather than transmutation. "I counsel you, O my Friends, that you do not work but about *Sol* [gold] and Luna [silver], reducing them into the first Matter, our *Sulphur* and *Argent vive:* therefore, Son, you are to use this venerable Matter; and I swear unto you and promise, that unless you take the *Argent vive* of these two, you go to the Practick as blind men without eyes of sense." [16] (He was clearly a disciple of Roger Bacon.)

Peter Bonus lived in the fourteenth century, possibly at Pola, a seaport on the Adriatic, and wrote an alchemistic work, the *Margarita Pretiosa* (of which an abridged English translation by Arthur Edward Waite was published in 1894). Peter Bonus agreed with earlier writers that all metals consist of mercury and sulfur, but arrived at his own conclusion that, while mercury was always the same, different metals contained different sulfurs. There was, he said, inward and outward sulfur, and to produce gold the outward and impure sulfur must be eliminated.

> Each metal differs from all the rest, and has a certain perfection and completion of its own; but none, except gold, has reached the highest degree of perfection of which it is capable. For all common metals there is a transient and a perfect state of inward completeness, and this perfect state they obtain either through the slow operation of Nature, or through the sudden transformatory power of our Stone. We must, however, add that the imperfect metals form part of the great plan and design of Nature, though they are in course of transformation into gold. For a large number of very useful and indispensable tools and utensils could not be provided at all if there were no copper, iron, tin, or lead, and if all metals were either silver or gold.[17]

Paracelsus

Through the end of the fifteenth century the alchemists cannot be regarded as having made substantial contributions to the advancement and practice of medicine, but they had paved the way for an erratic genius who pioneered in the fields of chemical pharmacology and pharmacotherapy. Theophrastus Bombastus von Hohenheim, or Paracelsus (1493–1541), has been hailed as "the most original medical thinker of the 16th century." [18]

"Alchemy is to make neither gold nor silver," Paracelsus wrote before he was twenty-five; "its use is to make the supreme essences and to direct them against diseases." [19]

Paracelsus did not of course arrive at this conclusion spontaneously. His first contact with metals, metallurgy, and alchemy had come early. He was born at Einsiedeln, near Zurich, Switzerland, where his father was physician to the town and its pilgrim hospital. ("Two hundred thousand pilgrims still worship the Black Virgin annually. They come by train today, but in the fifteenth century they followed the road that crosses the Sihl. They all passed the doctor's house, and many of them, weary and sick from long traveling, must have called on him." [20] But penniless pilgrims and the local woodchoppers could not support a doctor and his family. When Paracelsus was nine years old Wilhelm von Hohenheim became municipal physician of Villach in Karinthia, where he also taught "chemistry, or . . . alchemy in its progress toward chemistry" at the Bergschule, established by the Fuggers of Augsburg (who were engaged in mining activities in the vicinity of Villach) to train overseers for the mines and analysts to analyze the metals and ores discovered.[21] Dr. Wilhelm retained these posts until his death thirty-two years later, being then described in a document prepared by town officials as "the learned and famous Willhelm Bombast von Hohenheim, Licentiate of Medicine, [who] lived amongst us in Villach for thirty-two years and all the time of his residence led an honourable life and behavior." [22] The selection of Wilhelm von Hohenheim for the post at the school indicates that he was proficient in chemistry; it is likely that his son learned some of its principles from him and may already have conducted experiments of his own.

Between the ages of nine and sixteen Paracelsus attended both the Bergschule at Villach and a nearby Benedictine school at St. Andrew's monastery. This being the heartland of metal deposits, Paracelsus became well grounded in alchemy, mineralogy, and chemistry. Years later he would write in his *Chronicles of Karinthia:*

> At Bleiberg is a wonderful lead-ore which provides Germany, Pannonia, Turkey, and Italy with lead; at Hütenberg, iron-ore full of specially fine steel and much alum ore, also vitriol ore of strong degree; gold ore at St. Paternion; also zinc ore, a very rare metal not found elsewhere in Europe, rarer than the others; excellent cinnabar ore which is not without quicksilver, and others of the same character which cannot all be mentioned. And so the mountains of Karinthia are like a strong box which when opened with a key reveals great treasure.[23]

At sixteen Paracelsus studied for a time under Johannes Trithemius (1462–1516). Trithemius was a Benedictine monk, so erudite that at the age of twenty-one he was named abbot of Sponheim. In 1506 he was transferred to the monastery of St. Jacob, near Würzburg, and it was here that Paracelsus was enrolled among his students. While the abbot was a strict devotee of the Holy Scriptures and insisted that his pupils follow in his train, he nevertheless experimented successfully in magnetism, telepathy, magic, astrology, and alchemy.

The year 1510 found Paracelsus at the University of Basel. He seems soon to have become conscious that all he was getting was a dull reiteration of ancient formulas. When he was about twenty-two he went to work in the silver mines and laboratories of the Fügers at Schwatz in the Tyrol, about 30 kilometers from Innsbrück. (The Fügers of Schwatz are not to be confused with the Fuggers of Augsburg who mined Bleiberg.)

Paracelsus' work with the miners revealed to him the risks they had to run, the hardships they had to endure, and their occupational diseases. His work with the chemists, who were still in fact alchemists, convinced him of the futility of "gold-cooking." But he recognized that the combinations and solutions in their crucibles, retorts, and phials might be turned to the making of medicines—that all minerals subjected to analysis might yield curative and life-giving secrets. The combined experiences, according to an early authority, allowed him to enter into "the innermost recesses of nature" and to see through "the forms and faculties of metals and their origins with such incredible acumen as to cure disease." [24]

On the practical side, Paracelsus' experimental research in metals led him to the discovery of the medicinal value of chloride and sulfate of mercury, flower of sulfur, and possibly antimony. Zinc ointment, in use to this day, was developed during his stay at Schwatz. In medical practice he also made use of iron, lead, copper, potassium, and arsenic. On the hypothetical side, foreshadowing modern medical concepts, he produced the theory that the healthy human body is a particular combination of medical substances, with illness resulting from a disturbance of this combination. Consequently, illness could be cured only through the employment of chemical medicines. In taking this step, he broke with the teachings of Galen and Avicenna, the accredited authorities on matters medical, bringing himself into conflict with the medical establishment.

After Paracelsus

The teachings of Paracelsus required that a distinction be made between the chemists who pursued chemical studies with the purpose of discovering and preparing useful medicines and the alchemists whose objective remained the transmutation of base metals into gold. Sometimes, of course, a man might be both an alchemist and a chemist. Andreas Libavius (1540–1616), physician and teacher of Coburg and Halle, Germany, was a firm believer in the transmutation of metals, but actively directed himself toward the preparation of new and better medicines, and he is credited with having written the first textbook of chemistry. Jean Baptiste van Helmont (1577–1644) of Brussels, a follower of Galen who came to accept at least a part of the teachings of Paracelsus, was a prime mover in overthrowing the old medical doctrines. While van Helmont's chemical research greatly advanced the infant science, he remained a firm believer in the philosopher's stone and the transmutation of metals. The German Johann Rudolf Glauber (1604–1668), while dedicated to alchemy, devoted much of his attention to applied chemistry and developed Glauber's salt (crystallized sodium sulfate), which is used as a purgative to this day.

Beyond this differentiation between chemists and alchemists, there was division among the chemists. The followers of Paracelsus demanded that a *new* medicine, based on a philosophic concept of body chemistry, be substituted for much (if not all) of ancient medicine; the adherents of the Galenic tradition, while conceding that new remedies were useful and necessary for the physician, continued to believe that health was achieved by the perfect balance of the four humors—blood, phlegm, yellow bile, and black bile.

Paracelsus wrote prolifically, but little of his output was published until thirty years after his death. The appearance of his works produced considerable critical comment, pro and con. Basically the commentators disagreed as to the extent to which chemical medicines should be employed. In 1571 Peter Severinus (1542–1602), a student of the works of Paracelsus, produced a codification of his master's writings. It remained the recognized source for more than a century. The same year,

Johannes Guintherius of Andernach (1505–1574) attempted to integrate the views of the Paracelsians and the Galenists.

The Renaissance saw an upsurge in the use of metals in medicine; it was based on the new ideas and approaches provided by Paracelsus. In a sense the cause of the strict alchemists was a lost one. They never, so far as is known, transmuted a base metal into gold. But for their labors, however, Paracelsus and his immediate followers might never have developed the concept of chemical medicine—surely a spin-off of alchemy. Modern times have justified the faith if not the ends of the alchemists. Radioactive metals under bombardment are indeed "transmuted," not literally into gold but into a kind of "gold" that saves and prolongs life.

PART TWO

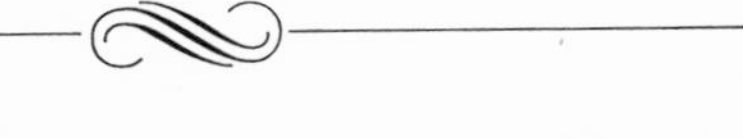

Emergence—The Fifteenth through the Nineteenth Century

5

The Triumphant Chariot of Antimony

It is to be recalled that, while antimony sulfide was known in ancient times, antimony was not then regarded as a metal. In point of fact, as late as 1677 the London Pharmacopœia was still drawing a distinction between *metals* (gold, silver, copper, tin, lead, and iron) and *"metallis affinia"* (cinnabar, mercury or argentum vivum, antimonium).[1] By one account, antimony was discovered by Basil Valentine, a Benedictine monk who lived at Erfurt, in what is now East Germany, in the fifteenth century. (Some authorities have placed his birth date as 1393.) The story goes that he tried out his new metal on some hogs. They fattened and exhibited unprecedented vigor and good health. Encouraged by this experiment, he fed his metal to his brother monks. The outcome was disastrous. What had proved good for hogs was bad for monks. Valentine named his metal anti-moine (anti-monk).

Basil Valentine is also credited with being the author of

the *Triumphant Chariot of Antimony,* first published in Leipzig in 1604. If he was, his writings must have circulated in manuscript for roughly two hundred years. But "no one has ever pretended to have seen one of those manuscripts." [2] There is in fact no certainty that Basil Valentine actually lived. A careful search of the provincial list and general role of the Benedictine monks, maintained by the order in Rome, failed to reveal a monk named Basil Valentine, and it is even questionable whether the Benedictines had a monastery at Erfurt.

Johann Thölde (fl. 1600) is recognized as the "editor" of Valentine's writings; it is highly possible that he is also their author. (It was not uncommon for writers of another day to attribute their books to someone of acknowledged fame, but it was unusual for an innovator to hide his light under an imaginary bushel of his own creation.) Some support for his authorship is offered by the fact that Thölde had, prior to publication of the *Triumphant Chariot,* produced under his own name a book in four parts—*Haliographia.* The fourth part was later included in the Valentine writings. "One of two things therefore is obvious. Either Thölde adopted a work of Valentine and issued it as his own, or one at least of the pieces alleged to have been written by Valentine was really by Thölde." [3]

There is, furthermore, intrinsic evidence that, if Basil Valentine was indeed the author of the *Triumphant Chariot,* he must have lived and written later than the fifteenth century.

The author tells us that the "Oyl of the *Mercury* of *Antimony* . . . heals the *French Disease,* which we have lately inherited; for by this Medicine it is radically exterpated." [4]

The "French disease" derived its name from the fact that it first appeared in epidemic proportions among the soldiers of Charles VIII of France at the siege of Naples in 1495. Involved was some form of venereal disease (possibly syphilis) that was communicated to the invaders by the Spanish occupants of Naples who in turn had contracted it from the sailors of Columbus returning from their voyage of discovery in the New World. To write of it, therefore, the author would have to be at least a contemporary of Paracelsus. (The picture of Basil Valentine in the Royal Cabinet of Etchings at Munich represents a monk who looks like the usual pictures of Paracelsus.) [5]

The foregoing discussion may seem as futile as the suggestion that the Shakespearean plays were not written by William Shakespeare but by a contemporary who went by the name of William Shakespeare, but a basic question is involved. Did Paracelsus, who contributed greatly to the launching of antimony's career in medicine, derive his knowledge from the writings of Valentine (it would have had to have been from one of the elusive manuscripts) or did Thölde (the pseudo-Valentine) borrow from the master—Paracelsus?

> It must be admitted that if Basil Valentine is a mythical character, the reputation of Paracelsus is greatly enhanced. Nowhere does the latter claim to have been the first to introduce antimony into medical practice, but it is certain that it could not have been used to any great extent before his time. If we suppose that the works attributed to Basil Valentine were fictitious, so far, that is, as their authorship is concerned, they were compiled about fifty years after the death of Paracelsus, and at the time when his fame was at its zenith. Many of the allusions to antimony contained in those treatises might have been collected from the traditions of the master's conversations and writings, much from his immediate disciples, and the whole skilfully blended by a literary artist.[6]

The Triumphant Chariot

The author of the *Triumphant Chariot* devotes fifty beginning pages, or about one-third of its content, to introducing himself in his pious surroundings and viewpoints, discussing chemistry at large, and finally upbraiding the doctors and apothecaries of the day.

And whensoever I shall have occasion to contend in the School with such a Doctor, who knows not how himself to prepare his own Medicines, but commits that Business to another, I am sure I shall obtain the Palm from her: for indeed that good Man knows not what Medicines he prescribes to the Sick; whether the Colour of them be white, black, grey or blew, he cannot tell; nor doth this wretched Man know, whether the Medicament he gives be dry or hot, cold or humid; but he only knows, that he found it so written in his Books, and thence pretends Possession (or as it were Possession) by Prescription of a very long time: yet he desires no further information. . . . A Paper Scrol in which their usual *Recipe* is written, serves their purpose to the full, which Bill being by some Apothecaries Boy or Servant received, he with great noyse thumps out of his Mortar every Medicine, and all the Health of the Sick.[7]

Finally, after a number of false starts, antimony was defined:

Antimony is a Mineral made of the Vapour of the Earth changed into Water, which Spiritual Syderal Transmutation is the true *Astrum* of *Antimony;* which Water, by the Stars first, afterward by the Element of Fire, which resides in the Element of Air, is extracted from the Elementary Earth, and by Coagulation formally changed into a tangible Essence, in which tangible Essence, (*viz.* whence *Antimony* is formally made) is found very much of *Sulphur* predominant, of *Mercury* not so much, and of *Salt* the least of all three; yet it assumes so much *Salt,* as it thence acquires an hard and immalleable Mass. The principal Quality of it is dry and hot, or rather burning, of Cold and Humidity it hath very little in it, as there is in Common Mercury; in Corporal Gold also is more Heat than Cold. These may suffice to be spoken of the Matter, and three Fundamental Principles of *Antimony,* how by the *Archeus* in the Element of Earth it is brought to perfection.[8]

Valentine-Thölde then devoted a number of pages to establishing that "*Antimony* is meer Venom, not of the kind of the least Venoms, but such, as by which you may destroy Men and Beasts." [9] He interrupted his discourse to discuss "the Process and Preparation of *Antimony;* for I little value the Clamours of arrogant and self applauding men: let them make and bring to light any Work that can excel *Antimony.* It is well known to me, that of *Antimony* may be made equal to Those, which are in Gold and vulgar Mercury (I except the *Astrum* of *Sol*) for of this may be prepared *Aurum potabile* against the Leprosie, of this may be made Spirit of Mercury, the highest Remedy against the French *POX,* of this other infinite Remedies may be prepared. If those Contemners cannot perceive and understand this, what wonder is it? None, because they have not learned it." [10] But he drifted through several more pages on venom before he was finally ready:

> Therefore now will I distinctly declare, how Medicine is to be prepared, Venom to be expelled, Fixation to be set about, and a true Separation to be made, by which the Evil may be subdued and depressed, and the Good triumph and be taken into use. [11]

This emphasis on antimony as a poison is interesting. Paracelsus successfully established the metal as a therapeutic agent, but some French physicians became convinced that it was a dangerous poison, and in 1566 the use of antimonials was prohibited by royal decree. (This edict was circumvented by prescribing white wine in cups of antimony or having patients swallow retrievable and reusable balls—perpetual pills—of antimony.) The *Triumphant Chariot* was published about forty years after the prohibition. Were the six or eight pages devoted to the poisonous aspects of antimony a defense against a ban already in force?

The author devoted about eighty (often rambling) pages to "how Medicine is to be prepared," etc., and the uses to which such medicine was to be put. He dealt with the metal itself (the regulus of antimony) and such compounds and derivatives as the oxysulfide (the glass) and a tincture made from it, an oil, an elixir, the argentine flowers of antimony (obtained by inflaming, volatilizing, and condensing the regulus), the liver of antimony (a compound containing a larger proportion of sulfide than did the glass), the white calx, and a balsam, among many. The regulus, in alloy with tin, was used to produce antimonial cups for the making of antimonial wine and the perpetual pills, "the surface of [which] became slightly oxidised, and consequently acquired a medicinal effect." [12]

Observations in the Eighteenth Century

Antimony was outlawed in Paris in 1566. Almost two hundred years later John Huxham (1692–1768) of Devon, England, a pupil of Hermann Boerhaave (1668–1738) of Leyden, who was regarded as the leading physician and teacher of his day, published *Medical and Chemical Observations upon Antimony,* which earned him the Copley medal. His observations began:

> Not above two Centuries ago, a Physician, who prescribed Antimonials, was expelled the Faculty; nay, at *Rome,* any one, who used Pulvis Cornachini, incurred the Penalty of being sent to the Gallies, on Account of the Antimonium Diaphoreticum, that was in it. Now, on the contrary, Antimony, in some Form or other, is the grand Catholicon, and used by Dabblers, as well as Doctors, in Physic. It is without all Doubt a most excellent Mineral, when duly prepared, and judiciously administered.[13]

Antimony had been returned to favor about a century before Huxham wrote. In 1657, Louis XIV, the nineteen-year-old king of France, became dangerously ill while at Calais. A physician (Voltaire says a quack) at Abbeville, about halfway back to Paris, treated him with emetic tartar (cream of tartar and argentine flowers of antimony—today known as antimony potassium tartrate). The king and his court were convinced that the monarch owed his life to this remedy. The opponents of antimony were silenced, and in 1666—just a hundred years after its promulgation—the decree against antimony was repealed.

In 1651 Glauber had devised an orange-red powder which he named Kermes mineral (later to be known as sulfurated antimony). Glauber kept his formula secret, but after his death in 1668 one of his pupils revealed it to a French physician named de Chastenay, who in turn passed it on to a surgeon named La Ligerie. Called *poudre de Chartres,* it became France's most popular remedy for many serious diseases, including smallpox, ague, dropsy, and syphilis. In 1720 Louis XV bought the formula from La Ligerie for a considerable sum. In 1751 an English physician, Andrew Plummer, published his formula for a golden powder. This powder was formed by adding hydrochloric acid to the residue of Glauber's solution. The further addition of mercurous chloride produced Plummer's pills. In 1747 Robert James (1703?–1776), a lifelong friend of Samuel Johnson and author of a massive dictionary of medicine and a *Pharmacopœia Universalis,* patented an antimonial fever powder, the use of which proved highly effective. A century after James's death the question was raised "whether his success depended upon his powder or on the mercurials and [cinchona] bark which he commonly employed at the same time." Patrick O'Connell, a Chicago physician, attempted to settle this question. Treating twenty-two gravely ill patients, he used James's powder, calomel (mercurous chloride), and opium on two; James's powder, quinine (the active element in cinchona), and opium on one; James's powder and quinine on three; James's powder and opium on three; and James's powder alone on thirteen. The thirteen cases included two (out of four) of intermittent fever, one (out of two) of puerperal pyrexia, one (out of two) of puerperal metritis, seven (out of eight) of acute pneumonia, all lobar, a case of pyrexia following postpartum hemorrhaging, and one of acute metritis and pelvic cellulitis, with effusion. (The other four cases involved acute peritonitis.) Dr. O'Connell reached this conclusion: "In these thirteen cases the action of James' powder alone was just as rapid, and the effect as complete and as satisfactory, as when combined with the quinine, calomel or opium." [14]

Huxham had cautioned that "whoever would give Antimonial Medicines with Safety and Success, should be well acquainted with the Analysis of that Mineral, and its component Principles; should know what different Combinations, Preparations, and Doses of them effect; otherwise it may prove a Poison, instead of a Remedy." [15] He then devoted much of his dissertation to expanding on and defining the chemical combinations and derivatives of antimony in use in his day, concluding that he did not

> pretend that the Observations I have here laid down are altogether new; I allow that far the greater Part of them are commonly known. . . . But, if I mistake not, I have, in some Measure, given a new Light into the Nature of Stibium [antimony]: at least have made it more obvious to the younger Part of such, whose Business it is to prepare and exhibit Antimonials, than they will readily find in any one single Treatise. And, as stibiate Medicines are now so much in Vogue, this little Piece may not be an improper Thing to put into the Hands of Students in Physic: . . . after all, it is not this or that Medicine, or Preparation, will cure a Disease, unless prudently made Use of. A Man may as perfectly well know how to make a Hatchet, a Hammer, or a Saw, as a Chemist how to make such or such particular Medicines; and yet the first may be as far from being a good Carpenter, as the second from being a good Physician: the Arcanum is how to use them.[16]

Huxham did not devote much space to the medical uses of antimony in its various forms. After discussing the efficacy of emetic tartar and the use of essence of antimony as a diaphoretic, he did, however, offer this summary:

> In obstinate Rheumatisms then, in cold scorbutic Affections, in most cutaneous Diseases, in asthmatic, leucophlegmatic, and icteric Disorders, in old stubborn Head-achs, Vertigo, Epilepsy, and Mania, Antimonials are very useful, and the Vinum Antimoniale in particular. In my own Practice I have had numerous Instances of its Success in the above Cases, and have likewise had the Pleasure in finding it sucessfully used by several eminent Practitioners.
>
> Let me further add, before I quit the Subject, that I very frequently give this Antimonial Wine, or Essence of Antimony, as I call it, in some acute as well as chronic Disorders, and particularly in slow Fevers, low irregular Intermittents and Remittents, in catarrhal Fevers, in a Peripneumonia Notha, and even in a true Peripneumony, after proper Evacuations, towards the Close, when the Spitting is prematurely suppressed, and great Anxiety and Difficulty of Breathing come on. In like Circumstances, it is very proper in the Small-pox also; and I have had the Satisfaction, through divine Goodness, of seeing it many Times very happily succeed in many desperate Cases; the Expectoration returning sometimes with a gentle Vomiting, sometimes a Stool or two, and sometimes a universal kindly Sweat.[17]

Toward the close of the eighteenth century William Blizard, surgeon to the London Hospital, decided to look into the external use of emetic tartar. He applied lint moistened with a saturated solution of emetic tartar to the surface of many ulcers. He found:

> 1st, It immediately occasioned a great degree of pain.
>
> 2dly, A florid appearance succeeded a foul aspect.
>
> 3dly, It constantly reduced the granulations in such a manner as generally, after the first and second application, to occasion a

> cavity in the ulcer. The effect seemed as is produced by an extraordinary action of the absorbent vessels, and not by the destruction of the living solids; for there appeared not the least sign of a dead surface, slough, or eschar; on the contrary, the face of every sore had constantly a red and healthy appearance, although continually wearing away.[18]

After discussing a variety of cases in which the external use of emetic tartar and proved effective, Blizard asked: "Do not these facts afford encouragement to make experiments with other combinations of antimony, and even with all the metallic salts, in many topical complaints?" [19]

Experiments involving antimony and particularly emetic tartar were undertaken sporadically during the nineteenth century but they were largely inconclusive. Today in the United States the use of antimonials, which have never lived down their venomous reputation, is a strictly limited one, but they are still employed extensively in tropical countries to combat parasitic infestations.

6

The Liquid Metal

The external use of mercury in the form of frictions, fumigations, and plasters for the cure of lues venerea (syphilis) dates from 1497. The famous Italian surgeon and anatomist Berengario da Carpi (1470–1530) of Bologna accumulated an immense fortune by inventing and prescribing friction with mercurial ointment for syphilis. A contemporary of Berengario, Giovanni de Vigo (1460–1520), born at Rapallo near Genoa and physician to Pope Julius II, was a staunch advocate of fumigations involving cinnabar and storax (tree resin). He may also have adopted the internal use of the red precipitate for syphilis; he clearly did so in cases of plague.

On the subject of early internal use, according to Gabriele Fallopio (1523–1562), Italian professor of anatomy, shepherds in his day gave quicksilver (the metal mercury) to their flocks to kill worms; Antonio Musa Brassavola (1500–1555) of Ferrara (who described two hundred kinds of syphilis) gave children doses of two to twenty grains that caused them to expel worms; Pietro Andrea Mattioli (1500–1577), Italian botanist and author of a number of books (including one on syphilis), told of a woman who took a pound of quicksilver to procure an abortion and suffered no ill effects. But whoever was the first to administer mercury internally, Paracelsus "was without doubt the practitioner who popularised its use. He gave red precipitate, corrosive sublimate, and nitrate of mercury, and described how each of these was made." The great Pomeranian medical historian Kurt Polycarp Sprengel (1766–1833) attributed to Paracelsus knowledge of mercurous chloride (calomel).[1]

William Clowes (1544–1604), probably the greatest English surgeon during the reign of Elizabeth I, published a treatise on lues venerea in 1585. It was based on the successful treatment over a five-year period of about a thousand patients at St. Bartholomew's Hospital. He gave numerous mercurial formulas for both external and internal use, including the combination of mercury and sulfur which the famous Sir Theodore Turquet de Mayerne (1573–1655) would subsequently name turbith mineral. Turquet de Mayerne, born in Geneva, taught and practiced in Paris but, when attacked by the Paris faculty for his free use of mercury and antimony, fled to London, where he became physician successively to James I and Charles I.

Thomas Dover (1660–1742), famous for his Dover's powder (opium and ipecac), a medicine for aches and pains that sells to this day, believed mercury in its crude state to be a valuable remedy for afflictions of the stomach and other diseases. "To take an ounce of quicksilver every morning, he declared to be the most beneficial thing in the world; and in 1731 and '32, it became 'fashionable' in London and Edinburgh to take that quantity every morning for several weeks." [2]

In 1735 there was an outbreak in New England of what would come to be known as ulcerated or malignant sore throat.

William Douglass (1692–1752), the only physician in Boston in his day with an M.D. degree, introduced a successful form of treatment involving dulcified mercury and camphor. But physicians generally were timid and capricious in following his lead, prescribing only small doses of mercury in various combinations. It remained for Jacob Ogden (1721?–1780) of Long Island, New York, who possessed more advanced pathological views and greater confidence than most, to establish in America the free use of mercurials in the treatment of inflammatory diseases.

At the beginning of the 1750s Edward Augustus Holyoke (1728–1829) of Salem, Massachusetts, began prescribing mercury in pleurisies and peripneumonias. In 1770 and 1771, when a disorder in many respects analogous to the malignant sore throat distemper appeared in the city of New York (and some other places), Samuel Bard (1742–1821), professor of medicine at King's College, New York, noting that the disorder was in general confined to children under ten years of age, decided upon the use of mercurials:

> The more frequently I have used them, the better effects I have seen from them. . . . [C]alomel is what I have commonly used, and have given it to the quantity of thirty or forty grains in five or six days, to a child of three or four years old, not only without ill effects, but to the manifest advantage of my patient; relieving the difficulty of breathing, and promoting the casting off the sloughs, beyond any other medicine.[3]

In 1781 Richard Bayley (1745–c. 1802) of New York was the first to recommend calomel in the treatment of hives. In 1783–84 Benjamin Rush (1745–1813) of Philadelphia liberally prescribed calomel in cases of scarlet fever with severe throat symptoms. A Delaware physician not only gave calomel in such cases but anointed the outside of the throat with mercurial ointment.

By the turn of the century mercury was being used extensively for diseases of warm climates, particularly malignant fevers.

The internal use of calomel had been accompanied by gastrointestinal symptoms so drastic that in the nineteenth century not only the chloride but most other forms of mercurial therapy fell into disfavor. The repudiation was led in part by a conservative Scotch surgeon, William Fergusson (1773–1846), who in 1813 made a comparison between mercury-treated syphilitic British soldiers and Portugese soldiers, who were not so treated and demonstrated fewer side effects. When the use of mercurials was gradually resumed, they were again denounced—this time by members of the New Vienna School of therapeutic nihilists, Josef Skoda (1805–1881), Josef Hamernjk (1810–1887), Josef Dietl (1804–1878), Carl Rokitansky (1804–1878), and Ferdinand von Hebra (1816–1880).

In 1857 the English syphilologist Henry Lee (1817–1898) suggested that calomel be employed in fumigation, but the efficacy of this form of treatment was already open to question. In 1861 Adolf Kussmaul (1822–1902) of Heidelberg attempted to absolve mercurial treatment from the evils attributed to it.

In 1839 Michael D. Donovan (1809–1876), Irish physician and pharmacist, invented a solution of arsenic and mercuric iodides. Prescribed originally as a tonic, its value in skin disorders was soon recognized. In 1919 Hugh Hampton Young (1870–1945), Baltimore urologist, produced (with E. C. White and E. O. Swartz) the great household standby—Mercurochrome.

It would not be fitting to leave this discussion of mercury without reference to several medically related instruments that employ the metal.

Having known nothing different, there is a tendency today for most people, in or out of medicine, to assume that the thermometer to be found in every home medicine chest was always activated by mercury, but such thermometers have been described as representing "a late step in a series of events which began some 2000 years ago." [4]

> Who invented or made the first mercury thermometers? Was it the Accademia del Cimento whose members tried and rejected mercury? Was it Boulliau? Was it Hooke? Or was it some unknown person? As in many claims for priority, national pride may threaten objective evaluation of available evidence. Perhaps the question will never be settled and, like the syphilis problem, remain as a never-ending source of material for professional as well as amateur historians. [5]

The sphygmomanometer is an instrument for measuring blood pressure that is based on a column of mercury. While the Reverend Stephen Hales (1677–1761) was apparently the first person to make an accurate measurement of blood pressure, the sphygmomanometer as it exists today was essentially developed by Scipione Riva-Rocci (1863–1936) in 1896. [6]

About 1912 Ernst L. F. Kromayer (1862–1933), German dermatologist, invented a water-cooled mercury vapor lamp provided with a quartz window that permitted the passage of ultraviolet rays.

7

The Noble Metals

Noble metals are distinguishable by the fact that they do not oxidize but remain bright when heated in the air, a condition that has led to their use for ornamentation. Consequently many of the noble metals are also classified as precious metals.

In the period under review gold and silver were the only noble metals employed medicinally, and their use was sometimes superstitious rather than scientific.

Royal touching to cure the king's evil (*morbus regius*), or scrofula (tuberculosis of the lymph nodes), dates back to the eleventh century. The practice was continued in France until the seventeenth century, but in England, while it was employed by Edward the Confessor shortly before his death in 1066, it fell into disuse under the Norman kings, only to be revived by the Plantagenets a century later. In 1492 Henry VII created an elaborate ritual that involved a special coin (the gold angel, minted in 1495) to be used as a touchpiece. The Stuarts favored touchpieces, little gold and silver medals to be hung about the necks of sufferers. Queen Anne even tried touching the young Samuel Johnson—but without success. (Royal touching ended in England in 1714 with the death of Queen Anne.)

> To be eligible for royal touching, patients were examined by physicians who certified that the complaint was scrofula. After being touched the patient received a gold piece, . . . and it was generally believed that he would be free from subsequent attacks provided he kept this "touchpiece." [1]

From the time of Edward II (1284–1327) to that of Mary Tudor (1516–1558), English sovereigns made a Good Friday offering of gold and silver. The metals were converted into "cramp" rings that were worn to cure epilepsy, rheumatism, and a variety of muscular pains.

Gold

Arnold of Villanova and Raymond Lully were loud in their praise of the medicinal value of gold, but in the first half of the sixteenth century Fallopio and Brassavola tried to discredit it. Paracelsus, however, prescribed gold as a purifier of the blood, as an antidote to poison, as a preventive for miscarriages, and, especially, as a cure for diseases connected with the heart, and with the support of Libavius and Oswald Croll (1580–1609), physician to Prince Christian of the German state of Anhalt, gold came into its own.

Francis Anthony (d. 1623), whose *aurum potabile* was famous during the reigns of Elizabeth I and James I of England, might have passed into history as just another quack had not the distinguished British chemist Robert Boyle (1627–1691) stated in his *Sceptical Chymist* (1661) that, while prejudiced against such concoctions, he had seen wonderful cures resulting from this *aurum potabile* and felt compelled to bear testimony to its efficacy.

Anthony claimed for his remedy the cure of a great variety of medical problems, including vomitings, fluxes, stoppages, fevers, plague, and palsy. When several prominent physicians of his day wrote pamphlets denouncing his claims, Anthony seized the opportunity to reply, thereby "advertising" and advancing his business. Because *aurum potabile* contained gold, it was immensely popular among the wealthy classes.

> The object, Anthony says, is so far to open the gold that its sulphur may become active. To open it a liquor and a salt are required, these together form the menstruum. The liquor was 3 pints of red wine vinegar distilled from a gallon; the salt was block tin burnt to ashes in an iron pan; these to be mixed and distilled again and again. Take one ounce of filed gold, and heat it in a crucible with white salt; take it out and grind the mixture; heat again; wash with water until no taste of salt is left; mix this with the menstruum, one ounce to the pint, digest, and evaporate to the consistence of honey. The Aurum Potabile was made by dissolving this in the spirit of wine.[2]

Other men produced other formulas. Around 1650 Glauber advocated "of living gold one part, and three parts of quick mercury, not of the vulgar." He allowed that a part of silver, equal to that of the gold, might be added but was not essential.[3] Kenelm Digby (1603–1665), "pirate for his Brittanic Majesty, commentator on poetry, gentleman of fortune, and *cognoscente,"* [4] promoted a powder of sympathy that cured wounds. His formula, which he claimed he had obtained from a Carmelite monk, involved "Gold calcined with three salts and ground with flowers of sulphur; burnt in a reverberatory furnace twelve times, and then digested with spirit of wine." Nicholas Lémery

(1645–1715), the self-taught French chemist and pharmacist who discovered iron in the ashes of animal tissues, followed this recipe: "Dissolve any quantity of gold you like in aqua regia; evaporate to dryness, and make a paste of the residue with essence of cannella. Then digest it in spirit." [5]

About 1540 a Paris physician, Antoine Lecoque (d. 1550), acquired a considerable reputation for his use of gold in curing syphilis. Fallopio, Friedrich Hoffmann, Jr. (1660–1742), and Archibald Pitcairn (1652–1713) of Edinburgh adopted his treatment to a greater or lesser degree, but the practice was gradually abandoned. It was revived in the latter part of the eighteenth century by Jean-André Chrestien (1758–1840) of Montpellier, who became an articulate advocate of the treatment.

According to candidate John C. Cheesman in his dissertation for the degree of Doctor of Physic (1812), the preparation Chrestien administered to his syphilitic patients came from an amalgam of gold and quicksilver. At first he employed the sun's rays directed through a lens to remove the mercury from the amalgam, but he subsequently substituted "common fire" as a vaporizing agent. In either case he was left with gold in the form of a "fine and subtile powder." He also made successful use of nitric acid, but was afraid that this method might leave small quantities of mercury adhering to the gold and that these might be credited with the cures. Thereafter he experimented with a number of agents in attempts to precipitate "oxyd of gold." They included potash, ammonia, and tin. Finally he combined the muriate of soda (sodium chloride—common salt) with the muriate (chloride) of gold "and obtained thereby the product he wished for" finding it "preferable to any other preparations, and possessed of infinitely more energy." Chrestien "for a long time employed all these preparations without giving his patients any other medicine whatsoever, that he might have indisputable evidence of their efficacy; and having become perfectly satisfied, he had now no apprehensions of using with them auxiliary means; though in venereal cases these are seldom necessary. So far from requiring collateral remedies to help them, he finds it necessary to calm and moderate the action of his muriate of gold, especially when prescribed for lues in irritable habits. This is accomplished by causing them to drink freely of whey." [6]

So far as Cheesman could ascertain, gold was first used medicinally in the United States in 1811 when the New York Hospital made a preparation that involved "taking a certain weight of gold, and dissolving it in nitro-muriatic acid, and evaporating the solution nearly to dryness; then adding to the salt thus obtained an equal quantity by weight of the muriate of soda, re-dissolve the mixture in a small quantity of water, and again evaporate by gentle heat in a sand bath as before: the residuum is the medicine here spoken of under the name of the Muriate of Gold." [7]

Cheesman considered the muriate of gold superior to mercurial preparations in the treatment of syphilis and other lymphatic affections for these reasons:

Mercury is idiosyncratic and can, even in small quantities, produce violent, injurious effects on the system.

Mercury, when administered in quantities sufficiently large to eradicate syphilis, frequently "produces symptoms of debility and disorder which often require much time and labour to relieve."

> To how many who are so unfortunate as to contract this disease, is it of the greatest consequence that it should be kept profoundly secret. This cannot with certainty be done if mercury is exhibited for its cure, there being no sure means of preventing salivation, let the Physician be ever so guarded in its administration. This occurrence is generally sufficient to cause suspicion, if not to disclose the nature of the disease. . . . [Gold] I am happy to say is not attended with any such unpleasant effects.

Valentine Seaman, a faculty member of Queen's College, New Jersey, a lecturer at the New York Hospital, to whom the dissertation is dedicated, had this to say:

> We have "indeed" seen mercury disrobed of half its honours, of its proud pre-eminence as the *only* remedy in syphilis; having now ample proof of the equal power of gold over that most dreadful disease.[8]

But how ample was Seaman's proof? Gold therapy in cases of syphilis had been practiced at the New York Hospital for only a year when Cheesman presented his dissertation. Furthermore, "trials at the French hospitals" made in response to Chrestien's advocacy "gave negative results." [9]

It appears that gold therapy never really caught on in America. In 1862 Samuel R. Percy, professor of materia medica and therapeutics, said in a lecture delivered at the New York Medical College and Charity Hospital: "Gold in a minute state of division has been used by many European physicians, but so far as my knowledge goes has been but little used in this country." [10]

Silver

In ancient times silver was associated with the moon, which was believed to rule the head. This relationship between the moon, silver, and the head was firmly believed in by the chemical doctors of the sixteenth century. Tincture of the moon (*tinctura lunæ*) was long a famous remedy for epilepsy, melancholia and mania. Angelus Sala (d. 1637) employed the nitrate of silver as a purgative and in cerebral and uterine diseases. Robert Boyle created *pilulæ lunares* involving a mixture of silver nitrate and nitre, dried and formed with bread crusts into pills about the size of small peas. Their purpose was laxative. Étienne-François Geoffroy (1672–1731) followed with a purgative that was identical with Boyle's. However, he fixed the dosage more precisely than Boyle had done. Hoffmann and Boerhaave made use of the tincture and the pills. Faith in the internal use of silver nitrate wavered as the eighteenth century advanced, but in a paper read at the Royal College of Physicians on 15 February 1808, Richard Powell (1766–1834), a fellow of the College, noted that there had been papers "in various modern journals" attesting to the internal use of silver nitrate, and added: "These make no mention of a purgative or other sensible effect supervening. They also, as far as I know, confine its use to epilepsy, a disease to which no remedy is generally applicable, since it arises from a variety of causes, and frequently from such as are beyond control." [11]

Powell went on to speak of "another convulsive disease" (*Chorea Sancti Viti,* Saint Vitus's dance) which could be treated with silver nitrate. Like epilepsy, it sprang from a variety of causes. "It may arise from sympathy with various local irritations, especially those of the intestinal canal, or from a particular state of the nervous and muscular systems, unconnected, as far as our powers of examination go, with any organ disease of the affected part." [12]

He reviewed some of the chorea cases in which he had used silver nitrate, noting that "I have had at one time within the last six months, no fewer than seven patients under my care in St. Bartholomew's Hospital, a larger number than Sydenham, according to his own account, had seen during the whole course of his practice." [13] (In 1686 Thomas Sydenham [1624–1689] had in his *Schedula Monitoria* focused some attention on chorea; in fact, a mild form of the disease is named after him.) From his experience, Powell inferred that

> we are not to expect from argentum nitratum those violent cathartic effects which the older writers have ascribed to it; that chorea, such as it occurs in this town, is by no means generally

> removable by purgative medicines, though in many cases they may still be obviously proper and effectual; that a remedy not in common use, and which had been considered dangerous, may be given with considerable freedom where the more common instruments of practice fail, and lastly, that this salt seems to produce a more decided effect upon morbid muscular contractions than any of the other metallic tonics.[14]

An early name for silver nitrate was lunar caustic, and fear that it might have a corrosive effect on the stomach inhibited its internal use. Even an advocate like Powell had to admit: "When I first began to administer the medicine, . . . I feared to proceed further than doses of one grain in a solid form, lest . . . it might act as a caustic upon the stomach." [15] By the third decade of the nineteenth century ingestion of silver nitrate had again fallen into disfavor. On the other hand its external use was gathering increasing support.

In 1826 John Higginbottom (1788–1876) of Nottingham, England, published a monograph on the use of lunar caustic in the cure of wounds and ulcers.[16] He recommended the free application of silver nitrate in the healing of wounds caused by needles, hooks, bayonets, and saws, for bites of leeches and rabid animals and stings of insects, and for small scratches and punctures occurring during anatomical dissection. In neglected cases, where tumors had formed under the skin, he advised removal followed by cauterization of the area with silver nitrate. In more advanced cases, he recommended that a cross-shaped incision be made before the caustic was applied and the area thereafter assuaged with a cold poultice and lotion.

Two years later Stephen Brown of the New York Hospital, while "giving full credit to Mr. Higginbottom's views of its efficacy," offered his own experience in the use of silver nitrate in "cases not mentioned by Mr. H." [17]

Brown claimed to have introduced successfully the use of

caustic in cases of inverted toenail (in substitution for the painful surgical removal of all or part of the nail), in the removal of corns, in ulcerations of the throat, mouth, and tongue, and in the relief of sore nipples.

Appended to the Brown article is a letter dated December 14, 1827, from a Dr. James Green to the editors of the *American Medical Recorder* reporting success in treating severe inflammation of the eye with silver nitrate. (In 1884 Carl Sigmund Franz Credé [1819–1892], German professor of obstetrics, instituted a method of preventing infantile [gonorrheal] conjunctivitis—*ophthalmia neonatorum*—by a single prophylactic instillation of a drop of 2 percent silver nitrate in each eye of all infants at the moment of birth.)

"After twenty years' further experience in the application of the nitrate of silver in the care of inflammation, wounds, and ulcers, I am desirous of giving full and clear directions for the use of it, particularly as the proper mode of application is quite essential to secure its good effect." [18] Thus begins a follow-up article by John Higginbottom written twenty-four years after the appearance of his small book. As a preface to this article he quotes Jacques Lisfranc (1790–1847), the inventive surgeon of La Pitié, Paris:

> Never is surgery so beautiful, and brilliant, as when obtaining a cure without the destruction of any organ, without plunging the bistoury [small surgical knife] into quivering flesh, and without causing the effusion of blood.[19]

Prejudice against lunar caustic had not abated.

> The great obstacle to the general and free use of the nitrate of silver, even at the present day, appears to arise from the impression in the minds of many surgeons that it is a caustic, a destructive agent. If they could be divested of that idea, and use it

as freely as they would a common blister of cantharides [dried insects obtained from Spain or Russia—Spanish fly], their fears would soon subside, from repeatedly observing the safety of the application, and also its beneficial effects. In my own practice I have always considered it a safer remedy than cantharides, as it may be applied freely over a surface, even when very active inflammation exists, or where there is an extensive surface denuded of its cuticle. This remedy has also the advantage of not affecting the bladder, or producing strangury [a slow, painful discharge of urine].

The nitrate of silver is not a caustic, in any sense of the word. It subdues inflammation, and induces resolution, and the healing process. It preserves, and does not destroy, the part to which it is applied. If we compare a caustic, as the hydrate of potassa [potash caustic] with the nitrate of silver, we find that the hydrate of potassa destroys and induces a slough, and the ulcerative process; but if we touch a part with the nitrate of silver, the eschar [slough] remains for a time, and then falls off, leaving the subsequent parts healed.

If an ulcerated surface, secreting pus, be touched by the nitrate of silver, the succeeding discharge is immediately converted into lymph; it is the property of the hydrate of potassa, on the contrary, to induce not only ulceration but suppuration. In short, the peculiar properties of the nitrate of silver have long been kept unknown to us by the designation of lunar caustic, affording the most striking instance of the influence of a term, or of a classification, upon the human mind. The nitrate of silver and the hydrate of potassa (as indeed all caustics,) are as the poles to each other; the first preserves, the second destroys; the first induces cicatrization [scar formation], the second ulceration.[20]

Higginbottom described his methods of silver nitrate treatment for recent bruised wounds of the shin, small and large ulcers, old ulcers of the legs, ulcers attended by inflammation, puncture wounds, bites and stings, wounds received in dissection, wounds from the bites of rabid animals, lacerated wounds, erysipelas and phlegmonous erysipelas, inflammation of the absorbents, phlegmonous inflammation, small irritable ulcers with varicose veins, senile gangrene, and burns and scalds.

The final category is of particular interest because it was not until 1965 that Carl A. Moyer of the Hartford Burn Unit (Washington University School of Medicine, St. Louis) "introduced" the silver nitrate treatment for burns, involving wet dressings of a ½ percent solution at frequent intervals.

Higginbottom's recommendation in the case of burns and scalds:

> In the first class of burns or scalds, where there is superficial inflammation, and in the second where there is simply vesication [blistering] without destruction of the cutis, the application of the nitrate of silver as directed in erysipelas, often effects a speedy cure; the vesicles should be removed, and the nitrate of silver applied on the exposed cutis, to form an eschar. . . . In burns from the explosion of gunpowder, particularly on the face, the mode of healing by eschar with the nitrate of silver is very successful.

Application "as directed in erysipelas" involved the following:

> Wash the affected parts well with soap-and-water, then with water alone, to remove any particle of soap remaining, afterwards wipe the parts dry with a soft cloth; then apply the concentrated solution of the nitrate of silver two or three times over the whole inflamed surface, and beyond it on the surrounding healthy skin, to the extent of two or three inches. In about twelve hours it will be seen if the solution has been well applied. If any inflamed part be unaffected by it, it must be immediately reapplied to it. . . . It is desirable to visit the patient every twelve hours, till the inflammation is subdued.[21]

By the final quarter of the nineteenth century the internal use of silver nitrate had returned to favor. In a paper read to the College of Physicians of Philadelphia on March 7, 1877, William Pepper (1843–1898), professor of clinical medicine in the University of Pennsylvania, spoke of

> the very numerous and varied conditions in which the internal administration of nitrate of silver is strongly indicated. Suffice it to say that there are few remedial agents whose therapeutic value and range of application are equally extensive. This is not surprising when it is remembered that nitrate of silver possesses a peculiarly valuable local action in addition to its powers of directly affecting the nervous system by absorption. The topical action of nitrate of silver, given internally, even in small doses, may be said to extend throughout the entire gastro-intestinal tract, with, perhaps, the exception of the rectum.[22]

Based on his personal experience, Pepper recommended the use of silver nitrate in numerous diseases:

> in the stomach—gastric ulcer, chronic catarrhal gastritis, especially with frequent vomiting or with pain dependent upon morbid irritability of the nerves of the mucous membrane; in the intestinal canal—cases of follicular (catarrhal) enteritis, both in the adult and in the child, subacute and chronic dysentery, chronic typhlitis, and tuberculous diarrhœa. I have lately obtained remarkable results from the administration of nitrate of silver in cases of gastro-hepatic catarrh, with or without jaundice from occlusion of the ducts by tumefaction of the mucous membrane. Recent observation has, moreover, led me to believe that the administration of this drug in typhoid fever, as originally recommended by [Jean-Christian-Marc-François-Joseph] Boudin [(1806–1867)] in 1836, is a very valuable element of treatment in this important affection. In addition to all of these must be mentioned the great value of injections of nitrate of silver in many forms of rectal disease.[23]

The nitrate was the form in which silver was mostly employed, but in 1849 C. H. B. Lane raised the question whether sedative action could be instituted through the agency of silver while avoiding the caustic stimulation of the nitrate, which frequently proved painful. His answer was the substitution of the oxide of silver for the nitrate.

After reviewing thirty cases (not all of them successful) in which silver oxide was employed "within my own knowledge during the last twelve months," Lane arrived at these conclusions:

> The oxide of silver is entirely devoid of causticity, its local application occasioning no pain, a valuable fact in reference to its internal administration. The remedy appears beneficial in various nervous affections, when they have become idiopathic, that is to say, when the cause, whether originally seated in the stomach, uterus, spinal cord, or other viscus, is removed, and the impression alone remains behind. There are no cases in which the oxide of silver is so rapidly beneficial as in cases of idiopathic gastric irritation, whether evinced in pyrosis, gastrodynia, or want of relation between the stimulus of food and the action of the stomach: but if organic change have taken place in the organ, the same benefit is not to be anticipated. In obstinate diarrhœa and hæmorrhages I am greatly in hopes that the silver will be found analogous in its action to lead—as efficacious, but milder and more manageable in its effect; this however requires much further trial. It would be unfair that the merits of the oxide of silver should be at all suffered to rest on its efficacy in epilepsy, of which we are well assured that the great majority of cases depend on organic change, which the medicine cannot influence.

Lane appended several written communications from physicians who had worked with silver oxide, including Golding Bird (1814–1854) of Norfolk, England, the author of the important *Urinary Deposits* (1844), who said:

> From the experience I have had in the administration of the oxide of silver, I have formed a high opinion of its value as a therapeutic agent; not from its possessing any marvellous specific

> power, but from its tonic, and to a certain extent, sedative properties—rendering it as far as I have seen a useful remedy in several forms of neuralgia, and especially in certain cases of dyspepsia attended with irritable stomach pain in the viscus after taking food, when the secretions of the liver and intestines have been corrected as much as possible by the careful administration of alteratives. The oxide of silver appears to me to possess the good qualities of the nitrate without its inconveniences, and to exert a more directly sedative action than that salt.[24]

In 1850 Francis Bennett of the Gateshead (England) Dispensary reported two cases in which he had avoided uterine hemorrhaging by the administration of silver oxide for three or four weeks after the seventh month. The more critical of the two cases involved a woman

> who had had previously five or six children, and in all her confinements there had been extensive hæmorrhage immediately after the birth of the child; in one only of these (the last) had I seen her. I must also mention that not only was she of the hæmorrhagic diathesis, (the slightest scratch causing the part to bleed for some time,) but that her sisters, who were married, and had children, suffered likewise from hæmorrhage.
>
> . . . At full period labour commenced and progressed favourably, although the doubts and anxiety were greater than before. Immediately before the expulsion of the head I gave her a dose of the ergot; a firm supporter was ready to be applied; the child was born; in fifteen minutes the placenta was expelled; and there was no more discharge than in the most common and favourable cases, and she very quickly recovered.[25]

The Tractors of Elisha Perkins

Was he a quack pure and simple or "an honest conscientious man, possessed of a strong character and a determined will; . . . full of faith in what he believed to be right"?[26]

Elisha Perkins (1741–1799) was eighteen when he began practicing medicine in Plainfield, Connecticut, in 1759. For almost forty years he was a respected physician, attaining the chairmanship of his county medical society. Then, in 1796, he patented "two metallic rods whose composition of dissimilar alloys was kept secret. They were gold and silver in color, rounded at one end, pointed at the other, and called 'tractors.' These tractors were alleged to relieve numerous painful ailments by the mere stroking of the afflicted parts." [27] (It was later determined that one of the alloys was gold, copper, and zinc and that the other was silver, platinum, and iron.)

The obtaining of this patent touched off a controversy that spread beyond the waters of the Atlantic. At home Perkins was eulogized by university professors, practicing physicians, and even the Chief Justice of the United States, but he was expelled by his state medical society. In England John Haygarth of Bath showed that like cures could be effected with wooden tractors—in short, that effectiveness lay in the imagination of the patient—but the royal physicians of Denmark attested to the value of the treatment, which they named Perkinism.

For thirty-five years Perkins had functioned as a typical backwoods doctor, riding as much as 40 miles a day to make housecalls. To supplement his income (and support his wife and ten children) he took in boarders from the Plainfield Academy and engaged in mule breeding and trading. Then in 1794 he achieved some prominence by developing an antiseptic for treating scarlet fever with severe throat involvement. It was a solution of sodium chloride and vinegar. "In retrospect, it would appear to have been effective because of astringent action, reducing edema and allowing for greater ease in swallowing foods and fluids." [28]

A year later Perkins was thinking in the direction of his tractors when he wrote his son-in-law, Joseph Arnold, that stroking a painful area with a polished knife blade would relieve the pain and that the then popular lacquered iron combs were even better for curing chronic headaches. He saw in his latest "invention" a means of benefiting mankind—as well as of relieving the near-poverty level of his existence. "If it was possible to monopolise the whole use of one hundreth part . . . it would make me and mine as rich as we ought to wish to be for it will, in a short time prevent the very great part of the pains that exist in the human machine." [29]

In 1771 the Swiss physician Franz Anton Mesmer (1753–1815) conceived of the magnetic properties of the human hand; in 1792 Luigi Galvani (1737–1798) of Bologna introduced the concept of therapy based on animal electricity; undoubtedly the tractors which Perkins patented in connection with his secret therapy were palpable tools to be used in the practice of animal magnetism and electricity, tools for which he could demand a substantial, even exorbitant, price.

While Perkins believed the tractors to be effective in practically all painful conditions, he proclaimed them particularly effective in gout, pleurisy, crippling rheumatism, "inflammatory tumors," and in calming and sedating violent cases of insanity. He also believed that tractor treatment, plus his antiseptic treatment, would cure yellow fever easily and quickly.

When yellow fever hit New York in July 1799 Perkins went to the city prepared to combat it. Within a matter of weeks he was dead from yellow fever.

> He went to New York confident in the ability of the tractors and antiseptic to cure the yellow fever. In this he was self-deluded perhaps, but no more quack or charlatan than the physicians who thought that it was the height and cold of the mountains that provided the effective principle in the treatment of tuberculosis.
>
> Perkins was a physician who displayed an interest in medical research and a desire to disseminate helpful information freely. Perkins, however, could not resist the temptation to emphasize the tractors in his treatment and obscure the value of the com-

moner knives and combs, to charge what appears to be an exorbitant price, and violate the medical traditions and ethics of his day. We can deny the charge that he was a quack and charlatan, for there is abundant evidence in his letters and in the manner of his death that he was a physician who sincerely believed in the efficacy of his method and the validity of his theories.[30]

8

The Semimetals

Certain elements are classified as semimetals because they possess much poorer conductivity than the true metals. One semimetal, antimony, has been dealt with separately because of its controversial history. Bismuth and arsenic have still to be considered.

Bismuth

Bismuth is ordinarily found in lead ores, but Lémery in the seventeenth century called it "a compound made in England from the gross and impure tin found in English mines," and the medical lexicographer John Quincy (d. 1722) claimed that it was "composed of tin, tartar, and arsenic, made in the northern parts of Germany, and from thence brought to England." [1] It is such promiscuous association with these and other metals that has led to confusion as to bismuth's efficacy in medicine. As late as 1818, for example, the French toxicologist Mathieu-Joseph-Bonaventure Orfila (1787–1853) was stressing the danger of arsenic poison from the internal administration of bismuth.

Georg Agricola (1490–1555), the German father of mineralogy, spoke of bismuth under that name in 1546, but it may have been known in Germany as early as 1470. Agricola regarded bismuth as a form of lead; some metallurgic chemists of his day believed that it gradually turned into silver. It was Georg Ernst Stahl who studied bismuth and established its characteristics.

Bismuth was first employed in salves in 1733, but while there may have been some internal use in Europe prior to 1786, fear of harmful effects inhibited investigation. Professor Louis Odier (1748–1817) of Geneva was not deterred, however, and in 1786 he made his findings public, indicating the disease entities against which the white oxide of bismuth could be applied with greatest advantage.

In July 1799 the editors of the *London Medical and Physical Journal* spoke of bismuth as a medicine either neglected or forgotten, notwithstanding the fact that it was considered a powerful remedy in spasmodic pain of the stomach and bowels. This neglect was attributable to fear of prescribing bismuth in large doses because it was frequently found in combination with lead or arsenic and might therefore prove toxic. Samuel A. Moore, in a dissertation delivered in 1810, referred to this situation and added: "I hope, nevertheless, that my readers can, at this time, entertain neither doubts nor fears on this subject." He described a method for preparing white oxide of bismuth which was "purer and whiter than that formed by any other process." [2]

In the meantime, Alexander Marcet (1770–1822) of Guy's Hospital, London, had delivered a paper which, when published in 1805, had brought the oxide of bismuth more generally to the notice of the profession. Speaking of "the white oxyd of Bismuth, commonly known by the name of Magistery of Bismuth, and sold chiefly to perfumers as a paint for whitening the complexion," he stated that its medical properties "are yet but little known, and have never, I believe, been submitted in this country to any regular investigation." Marcet continued: "Being at Geneva about twelve months ago, I heard that this substance had for many years past been brought into medical use by Dr. Odier professor of physick in that town, and employed there with considerable success, not only by him, but also by several of his colleagues in the treatment of a few spasmodic disorders, and more especially in the cure of a particular symptom of dyspepsia." [3]

Marcet quoted Odier on the preparation of the oxide:

> The magistery of bismuth is prepared by dissolving a quantity of very pure bismuth in nitric acid, and precipitating it by water, or by a solution of potash. But if the bismuth is not very pure, if for instance it is mixed with nickel, the precipitate is not perfectly white; it is then mixed with a greenish precipitate, which is nothing but an oxyd of nickel which water will not precipitate; for which reason we are more certain of obtaining a pure precipitate of bismuth by water than by potash. [4]

Marcet detailed six cases which he had undertaken at the turn of the century, concluding:

> If it be permitted to draw any inference from so small a number of trials, it would appear that the oxyd of bismuth is a remarkably successful medicine in spasmodic affections of the stomach; for in four cases out of six, in which it was tried, a complete cure was almost immediately obtained; and in the two instances in which it failed, the affection, which was at first suspected to depend upon a spasm of the stomach, has since appeared to be of a complicated, and probably, of a very different nature. [5]

In the printed version of his observations, Marcet added: "Since the above paper was read before the Society, I have had frequent opportunities, at Guy's Hospital, of trying the oxyd of bismuth in spasmodic affections of the stomach, and those trials have fully confirmed the opinion which I offered three years ago, on the utility of this medicine." [6]

Marcet's strong recommendation was taken up by Samuel Argent Bardsley of the Manchester Infirmary. In 1807, in *Medical Reports,* he had this to say on the effects of the white oxide of bismuth:

> In pyrosis [burning sensation], cardialgia [heartburn], and other local affections of the stomach the oxide of bismuth seems well calculated to afford relief. . . .
>
> It may be proper to mention that the oxide of bismuth is justly entitled to the attention of practitioners, on account of its safety as well as utility. For in no one instance did I find it prove injurious to the stomach or general system; nor as a medicine was it disgusting to the palate. . . .
>
> Since the above reports were sent to the press I have treated five cases of pyrosis, accompanied more or less with spasmodic pains of the stomach, with uniform success. In all these instances the bismuth, with occasional aperients, was solely employed.[7]

But fear, and its accompanying opposition, died hard. In 1845 I. P. Garvin, professor of materia medica at the Medical College of Georgia, still found that bismuth needed championing:

> The utility of the Sub-nitrate of Bismuth in certain painful affections of the stomach, has been known to the profession, ever since the publication of Odier, of Geneva, who was the first to employ it internally. . . . Our sole object is to invite attention to a most valuable remedy which we think is too much neglected. . . . Considerable fear is entertained by some lest poi-

sonous effects should follow the use of bismuth. It is true, that when imperfectly prepared, it may contain a small portion of arsenic in the form of an arseniate of bismuth, and to the presence of this substance must any ill consequences be attributed which may follow ordinary doses, for when the sub-nitrate has been prepared from the pure metal, precipitated and well washed, no danger need be apprehended.[8]

Garvin attributed a part of the problem to a lack of understanding of the action of bismuth. He quoted the French clinician Armand Trousseau (1801–1867) to this effect:

If we endeavor to ascertain the action of the sub-nitrate of bismuth, we will be much embarrassed; no intermediate effect between the employment of the medicine, and its curative results can be perceived. Notwithstanding the attention we have given to it, we have not been able to perceive the least influence on the general functions. When an individual in good health takes the sub-nitrate of bismuth, the only phenomenon to be noticed is constipation, but the nervous functions, the animal heat, the movements of the heart, the urinary and cutaneous secretions, are not influenced in an appreciable manner.

To which Garvin added: "We can therefore only infer the nature of its action, from the character of the derangements in which it operates beneficially." [9]

This empirical approach produced the information that, in addition to nervous derangements of the stomach, bismuth was beneficial in the vomitings of teething children, in the frequent diarrheas of feeble infants, in diarrheas unaccompanied by fevers following acute disorders, and like applications. In 1868 Adolf Kussmaul was the first to treat gastric ulcer with large doses of bismuth.

Salts of bismuth are still used to soothe the irritated membrane of the gastrointestinal tract. (TV viewers will be familiar with Pepto-Bismol.) These bismuth preparations supposedly coat the membrane and act mechanically as a protective. Bismuth compounds were, for a time, used in the treatment of syphilis but have been replaced by penicillin; however, they are still occasionally employed as a preliminary on syphilitic patients with gummatous laryngitis to decrease the likelihood of inflammatory reaction in the larynx (that could cause death) which might occur if penicillin were given first.

In 1898 the American physiologist Walter Bradford Cannon (1871–1945) introduced the bismuth meal in radiographic procedures; in 1908 Emil G. Beck, a Chicago surgeon, devised a bismuth subnitrate paste for the diagnosis and treatment of tuberculous sinuses and cavities and fistulous tracts; in 1916 James Rutherford Morison (1853–1939), an English surgeon, developed *bipp* (bismuth-iodiform-paraffin paste) for the treatment of wounds.

Arsenic

> Arsenic in solution—the aqua Toffana, so named after Toffa, a female poisoner executed at Naples in 1709, was such a solution—was prepared by Thomas Fowler (1736–1801) and employed (after the example of Dioscorides, Caelius Aurelianus and the Arabians—Marx) by Slevogt (1700), Fowler (1786), Adair (1784) and others.[10]

Arsenic had been used against syphilis in the fumigations of the sixteenth century. A Swiss physician, François Plater (1645–1711), employed, internally, a solution of arsenic and corrosive sublimate. But it was Thomas Fowler of Staffordshire, England, who was hailed at the close of the eighteenth century as "the only author who has treated systematically of the virtues of Arsenic." [11]

While work on an arsenic solution had been under way for some years, it was in 1786 that Fowler published in London his

Medical Reports on the Effects of Arsenic in the Cure of Agues, Remittent Fevers, and Periodic Headaches. It drew an immediate and favorable response in a letter to Fowler from William Withering (1741–1799), famous pioneer in the use of digitalis (from foxglove), written from Birmingham in May 1786:

> The *Arsenical Solution* was first used here in the autumn of the year 1783, at which time *intermittents* of various denominations were very prevalent both in this and in the adjoining counties. The very general use of the *Tasteless Ague and Fever Drops* at this period made me solicitous to know the composition of that medicine, and I was informed that it was made from the ore of cobalt; but as it did not answer to the usual tests for that semi-metal, I thought it was probably arsenic, which is known to abound very generously in cobalt ores. Whilst I was intending to submit the drops to more effectual examination, Mr. Hughes informed me that he had made an analysis of the drops, and found them to be a solution of arsenic.
>
> At first we gave the medicine cautiously, and were sparing in the doses; but it nevertheless gained our confidence, and in the vernal intermittents of the following year it came into pretty general use. Out of forty-eight patients, thirty-three were cured by the use of the solution alone; three complained of pain in the stomach, loss of appetite, and had swollen faces; but their fevers were cured, and a little soluble tartar removed the symptoms now mentioned. The other twelve patients received no benefit. In the autumn of 1784, it was almost constantly prescribed, and has ever since maintained its credit with us under a very great number of trials. . . . I have prescribed the medicine, as well in private as in public practice [and] I do not recall a single instance in which it produced either sickness, purging, pain, swelling, or any other than *curative* effects. Mr. J. Freer, a very eminent surgeon in Birmingham, who was one of the first to turn his attention to this medicine, tells me he has given it to more than a thousand patients, without either hazard or inconvenience. I think mankind much indebted to your endeavours to rescue a very useful medicine from the oblivion to which the general

> abuse of it would soon have consigned it; and I am satisfied that in patients where great debility has prevailed, either from old age or other causes, and where the recurrence of the fever under the quotidian form, with the long protraction of the paroxysms and the great irritability of the stomach, has not allowed a sufficient quantity of [Peruvian] bark to be given, I have seen their existence preserved by the use of arsenic.[12]

In 1842 Boudin, physician-in-chief at the du Roull French military hospital, who six years earlier had recommended the use of silver nitrate in the treatment of typhoid fever, reported on 264 fever cases he had treated with arsenic. Of these, 188 were cured by arsenic alone, 57 resisted Peruvian bark (cinchona) and were cured by arsenic, 11 resisted arsenic and were cured by bark, and 8 resisted both bark and arsenic.

Fowler limited the use of his solution of potassium arsenite to the treatment of fevers and headaches. In 1789 James M. Adair, in *Medical Commentaries of a Society of London Physicians,* outlined favorable results of arsenic administered internally in obstinate cases of such skin diseases as herpes, ringworm, and eczema. By one account Benjamin Rush reached similar conclusions at "about the same time," but Nathaniel Potter, a pupil of Rush, placed his participation somewhat later: "On my return to Philadelphia in the summer of 1792, I informed my preceptor, Dr. Rush, of what I had heard and seen concerning the internal use of arsenic; he soon had an opportunity of trying its efficacy in an obstinate herpetic eruption which had resisted the usual remedies." [13]

The French dermatologist P.-L.-Alphée Cazenave (1795–1877) wrote in the *Dictionary of Medicine* (Paris, 1844):

> It is now proven that wonderful results are obtained with arsenious acid in the treatment of cutaneous diseases, both in the

dry forms and in chronic eczema and intertrigo. This remedy is less successful in papulous eruptions, and in general it has almost failed in the various forms of porrigo, acne, and sycosis. It may be useful in the elephantiasis of the Greeks; to the treatment of acute exanthemata it is not applicable, as a general rule.[14]

Rush's student Potter recorded some earlier uses of arsenic:

In the year 1783, an itinerant practitioner, who called himself Lafferti, travelled through the state of Maryland: he astonished the practitioners of that country by curing ulcers long deemed beyond the reach of the surgical art, and it is not to be controverted that his success was unparalleled. It was observable, that he refused to undertake the cure of recent ulcers, and, unlike most of his empyrical brethren, candidly acknowledged that he had no skill in such cases as others considered most curable. The author's deceased father who at that time practiced physic and surgery in that country, left the following account of this practitioner, in a letter which he intended to have sent to a medical friend. "We have in this country (Caroline) a man who does wonders in the cure of obstinate old sores; but he uses so much mystery, and applies his powder with so much secrecy, that he does not seem to intend to let us into the secret. However, I have just procured a small parcel of his medicine; at first I thought it looked like corrosive sublimate, but upon trial found myself mistaken; I put some of it on the fire, which soon perfumed the room with the smell of garlic, from which it must be arsenic." Mr. Justamond, who has recommended arsenic so strenuously in cancer, mentions the authority of Sir Hans Sloane for its good effects in scrophulous ulcers, and thinks it a valuable medicine in such cases.[15]

At about the same time an American physician, Hugh Martin, developed an arsenic cancer powder which was to become celebrated and which caught the attention of Benjamin Rush in 1784:

> The great art of applying [arsenic] successfully, is to dilute and mix it in such a manner as to mitigate the violence of its action. Dr. Martin's composition was happily calculated for this purpose. . . . It excited a moderate inflammation, which separated the morbid from the sound parts, and promoted a plentiful afflux of humours to the sore during its application. It seldom produced an escar; hence it insinuated itself into the deepest recesses of the cancers, and frequently separated those fibres in an unbroken state which are generally called the roots of the cancer.[16]

In the eighteenth and early nineteenth century there was considerable opposition on the part of medical men to the use of arsenic, internally or externally, because it was a poison. Perhaps the best answer to such skeptics was given by William Withering when he said in *A Botanical Arrangement of All Vegetables Naturally Grown in Great Britain According to the System of the Celebrated Linnaeus* (1766): "Poisons in small doses are the best medicines; and the best medicines in too large doses are poisonous." To which Potter added:

> There is no substance so replete with death as not to admit of a safe application to the body, both externally and internally, by a certain degree of division or dilution; neither is there any so innocent as not to be capable of abuse by excess in quantity, or by long continued application. Opium, in portions accommodated to the excitability of the system, proves a most exhilarating cordial, and by its universal stimulus imparts tone to every fibre of the body; but increased to the quantity of a few grains, it dissipates life like a vapour. The oxygenated muriate of mercury, in doses of the fourth or sixth part of a grain, is an inoffensive and efficacious medicine, but ten grains of the same salt will annihilate life with irresistible impetuosity. Stramonium [Jamestown weed], digitalis purpurea, and cicuta virosa [hemlock], and many of the metallic preparations are invaluable medicines while they are confined within their proper limits, but in large doses

derange and disorganize the most vital functions of the animal economy. The virus of the spider, which has so often extinguished the lamp of life, when taken into the stomach diffuses the animating sensation of the mildest stimulant, and is often used in cure of intermittents with the most propitious effect.[17]

A nonmetallic element may be incorporated with a metallic element to produce a valuable compound. Iodine provides an instructive example.

Iodine-containing substances, such as seaweed and burnt sponge, had long been in use for thyroid and tubercular swellings, but they were not systematically exploited except in locations where goiter is common. That these substances depend for their action upon iodine was not realized until 1811, when iodine was discovered in *varec* (seaweed ash) by Bernard Courtois (1777–1838), a manufacturer of saltpeter. Courtois' discovery was announced to the French Institute on November 29, 1813.[18]

In 1838 Anthony Todd Thomson (1778–1849), compiler of the *London Dispensatory* (1824), read a paper to the Medical Section of the British Association for the Advancement of Science entitled *Preparation and Nature of the Iodide of Arsenic:*

Since the introduction of iodine as a therapeutic agent some of its compounds have been employed in medicine, and the iodides of iron, of mercury, and of potassium, now hold a place among the preparations of the London Pharmacopœia. The iodides of lead, of arsenic, of zinc, of sulphur, have also been medically employed; but little is known respecting their physiological action, and still less regarding their poisonous properties. . . . With a conviction of the value of one of them as a medicine of great power, I propose to lay before [you] my observations and practical experience of the medical properties of the iodide of arsenic, and an account of some experiments with it, as a poison, on dogs.

After reporting on patients who had been successfully treated and his experimental work on dogs, Thomson arrived at these conclusions:

> 1. That the iodide of arsenic, administered in non-poisonous doses, is carried into the circulation, and excites generally the capillary system, augmenting the energy of the secreting function, and operating as a powerful and beneficial alterative.
>
> 2. That it enters all the secretions; consequently, when therapeutically employed, it may be administered to infants at the breast through the milk of the mother.
>
> 3. That in moderate doses, not too long continued, it improves both the digestive and the assimilating functions; invigorates the muscular power, and augments the bulk of the body.
>
> 4. That the same salutary effects are obtained from its employment, as a therapeutic agent, whether it be administered in the solid form or in solution.
>
> 5. That when its internal use is long continued, it accumulates in the system, and displays symptoms of poisoning, namely, pain at the epigastrium and in the lower bowels, tremors, and general febrile excitement.
>
> 6. That, in large doses, it is a most virulent poison, inflaming the tissues to which it is applied, and softening and gelatinising the mucous membrane of the stomach, and, occasionally, causing ulceration.
>
> 7. That, when administered in large doses or applied to mucous or serous surfaces, or to an ulcer or a wound, it is absorbed, and exerts a deleterious influence on the nervous centres and the heart.
>
> 8. From its effects when its solution is injected into the veins, and from both sides of the heart being found turgid with coagulated blood, when death ensues from large doses introduced into the stomach, we may venture to conclude that, independent of its topical action, it kills chiefly by its powers of destroying the irritability of the heart.[19]

From 1846 to 1849 Thomas Hunt, an English surgeon, conducted a survey on the medicinal action of arsenic for the

Provincial Medical and Surgical Association. Seventy-five doctors responded, covering at least three thousand cases in which medical doses of arsenic had been administered internally. (Hunt excluded figures supplied by James Startin, founder of the London Cutaneous Institution, because Startin had observed "not less than *twenty thousand cases*" in which "he seldom or never gives the mineral *uncombined,*" while the seventy-five other respondents *"have,* for the most part, given it *uncombined."* To include the Startin cases would "deprive the Report of its value as a statistical document.")[20]

Hunt reached these major conclusions: Arsenic is a safe medicine, no respondent having found it either fatal or permanently detrimental to health; arsenic is medically efficacious, particularly in cutaneous affections, neuralgia, and intermittents and malarious disorders; because of its toxic character, arsenic should not be administered in cases in which less dangerous medicines are likely to prove equally effective.[21]

By 1858 James Begbie (1798–1869), vice-president of the Royal College of Physicians of Edinburgh and Physician in Ordinary to the Queen in Scotland, could report that

> much of the dread and apprehension of [arsenic's] poisonous effects, which deterred many from its use altogether, or suffered its exhibition to be so limited as to cause it to fall short of the production of its physiological operation, have happily been overcome; and the profession now generally regard it as one of the most useful and available of its therapeutic agents, powerful in many intractable affections, and exercising a commanding influence over ailments hitherto considered incurable.[22]

By and large the use of arsenic has been limited to fevers, skin diseases, and neuralgia, although the Hunt survey turned up a few successes in uterine affections (including hemorrhaging), chorea, chronic rheumatism, carcinoma, and dyspepsia, and single instances of use in a variety of diseases. In 1878 Steward Lockie, senior physician at the Cumberland Infirmary in the north of England, endorsed an extension of its use to blood and heart problems:

> I desire . . . to call attention . . . to the really remarkable results obtained by the administration of arsenic in certain cases of anæmia, and these cases in which iron and good food have failed to produce any benefit. . . .
>
> I have dwelt upon the value of arsenic in anæmia, as there its action as a blood-tonic is most strikingly seen, and because its use in these cases is probably less generally known; but it may well be that it is as a blood-tonic that it proves useful in chorea and in phthisis, both ailments associated with an anæmic condition.
>
> I pass on to a very brief consideration of the use of arsenic as a cardiac tonic.
>
> In 1871, the late Dr. Anstie . . . published a remarkable paper on the Pathological and Therapeutical Relations of Asthma, Angina Pectoris, and Gastralgia. In this paper, he says: "There is one remedy that is supremely effective, where it can be tolerated, in all these three maladies; namely, arsenic." Now, it is this element of pain in cardiac disease which has been usually, though not exclusively, looked upon as the indication for administering this drug, and there is a concurrence of experience as to its value in this respect. But I am convinced that other indications may be fulfilled by the use of arsenic. I believe it to be a valuable tonic to the cardiac muscle, and that it forms an useful adjunct to digitalis in ordinary vascular disease of the heart, where there is failure of compensation, with its consequent results. Further, it seems to be of great value in fatty degeneration.[23]

Arsenic as a specific for consumption (tuberculosis) was reported on by doctors as widely dispersed as Canada, New York, and southern Italy. In the winter of 1864–65 a series of articles appeared in the *Quebec Gazette* by a writer who "maintained that arsenic was a powerful remedy against pulmonary consumption, and stated that he himself had used it as such, with good effect, for many years, and was still in the habit of doing so from time to time." The case was investigated by F. A. H. LaRue of Laval University (Quebec). The writer proved to be an arsenic-eater. (About a year earlier Craig Maclagan of Edinburgh had visited Styria—a province of southeast Austria—and reported that it was not uncommon for peasants to eat the white trioxide of arsenic.) LaRue's "patient," now forty-seven, was in his early thirties when he believed he was attacked by hereditary consumption. (His father had died of tuberculosis of the lungs at the age of thirty-nine.) Having read somewhere that arsenic was an excellent remedy for consumption, he began both eating it and smoking it (mixed into his tobacco). He told LaRue: "I have read all the doctors have to say about arsenic and feel convinced that they know nothing at all about the matter." Firmly of the opinion that he would long since have died of consumption if he had not resorted to arsenic, he always carried the white powder with him, taking or smoking it whenever he developed a cold. (Under no circumstances would he take arsenic in solution.) [24]

In 1869 Iretus Greene Cardner of New York expressed his belief that the use of arsenic in consumption was "based upon the same general principle of its use in other diseases—its power of arresting disorganization, and of increasing the assimilative and reparative processes. Further than this general effect on the vital processes of the capillaries, it probably has no other effect on the lung tissue." [25]

For four years, 1872–76, Horatio R. Storer, acting on behalf of the American Medical Association, investigated the claims of health resorts in central and southern Europe. A frequently heard objection to resorts in southern Italy was the changeable winter climate, particularly unsuited to chest affections. Even in the more sheltered and more accessible spots, such as Pozzuoli, near Naples, accommodations suitable for invalids were unlikely to be available.

During 1873 and 1874 the writer's attention was chiefly given to the ordinary considerations of local climate, and a study of the Neapolitan chain of mineral springs extending from Meta, adjoining Sorrento, through the whole circuit of the gulf. . . . He then became aware, and he thinks it was from a chance statement of his friend, Dr. J. A. Menzies, of Naples, that several of the more noted Neapolitan physicians were commencing to claim for Pozzuoli an exceptional excellence of a wholly different character, asserting that a portion of it had an atmosphere of its own, perceptibly charged, not with sulphur merely, but with arsenic from the semi-extinct volcanic crater known as the Solfatara, which, from but slightly rising above the level of the adjoining country, is accessible on foot, or by donkey or sedan chair, to the most feeble invalid. The breathing of this arsenical atmosphere, it was stated, not only theoretically promised to be of benefit in cases of threatened or actual pulmonary tuberculosis, after repeated visits to the crater, extending over a longer or shorter period, but had in fact been proved so by actual experiments.

Storer was somewhat reserved in his recommendation to the association, partly because of "the total unfitness of Pozzuoli for the residence of invalids who had been accustomed to average American and English comforts." Still, he was of the opinion "that the arsenical and other emanations given out by the still smoking crater sensibly and beneficially modify the neighboring atmosphere in a medical way." [26]

Between 1906, when Wassermann announced his diagnostic test for syphilis, and 1910, the noted German researcher Paul Ehrlich (1854–1915) and his co-workers diligently searched for a drug that would sterilize the syphilis parasite. After 605 compounds had been investigated and discarded, they hit upon "606"—arsphenamine, or Salvarsan. In developing it, Ehrlich followed the maxim that a chemical substance is effective only on tissues by which it is fixed. Arsphenamine did not, unfortunately, reach all the spirochetes which, like the gonococcus, hide in other tissues, but it was nonetheless as reliable a specific as quinine in malaria and a valuable prophylactic in that it quickly cleaned up luetic sores and eruptions and sterilized blood, thus minimizing the possibility of infecting others. (It was not realized until some years later that for sterilization in syphilis a period of continuous arsenotherapy was necessary.)

Disappointment with arsphenamine led Ehrlich and others to search for superior arsenical antiluetics, and by 1922 four others (including neoarsphenamine—"914" in the Ehrlich experimental series) were in clinical use. Research in the arsenical field remained active until 1945, when penicillin was introduced into syphilotherapy. The only noteworthy present-day use of arsenicals in human medicine is in the treatment of trypanosomiasis (sleeping sickness) and amebic diseases.

9

The Common Metals

The common metals employed in medicine between the fifteenth and nineteenth centuries were zinc, magnesium, tin, iron, lead, manganese, and nickel.

Zinc

Erasmus Ebener of Nuremberg is credited with being the first to recognize zinc in its metallic form in 1509. Paracelsus named it *zinken* and offered this description of it:

> There is another metal, zinc, which is in general unknown. It is a distinct metal of a different origin, though adulterated with many other metals. It can be melted, for it consists of three fluid principles, but it is not malleable. In its colour it is unlike all others, and does not grow in the same manner; but with its *ul-*

tima materia I am as yet unacquainted, for it is almost as strange in its properties as argentum vivum [quicksilver].[1]

There was some small medical application of zinc and its salts in the centuries that followed. But while the presence of zinc in living organisms and its role as an essential nutrient for plants and animals was recognized in 1869, "only in recent years had it become possible to discern the presumable manner in which the metal participates in metabolism."[2]

Part of the reason for this neglect lay in differences of opinion about what zinc was.

> The ideas entertained of zinc by the chemists who studied it were curious. Albertus Magnus held that it was a compound with iron; Paracelsus leaned to the idea that it was copper in an altered form; Kunckel fancied it was congealed mercury; Schluttn thought it was tin rendered fragile by combination with some sulphur; Lemery supposed it was a form of bismuth; Stahl held that brass was a combination of copper with an earth [zinc] and phlogiston [fire]; Libavius (1597) described zinc as a peculiar kind of tin.[3]

By the eighteenth century zinc was creeping into the pharmacopoeias, possibly at the instance of Hieronymus David Gaubius (1705–1780), a pupil of Boerhaave, who became professor of chemistry at Leyden, He introduced oxide of zinc as an internal medicine. (It had in fact been known and used under the name of flowers of zinc for a century.) A Dutch shoemaker named Ludemann acquired something of a reputation through the sale of a medicine for epilepsy called *luna fixata.* Gaubius analyzed it and found it to be oxide of zinc. Gaubius did not endorse its alleged effectiveness in epilepsy, but in 1830 Ernst Klokow (1802–1867) was recommending hydrocyanate of zinc in the treatment of that affliction.

Nicholas Myrepsus (fl. 1275), an apothecary of the Byzantine city of Nicea, had produced an ointment, valuable in the treatment of malignant ulcers, that was composed of oxide of zinc, white lead, the juice of nightshade berries, and frankincense. In 1866, Campbell De Morgan (1813–1876), surgeon to the Middlesex Hospital, London, reported the use of applications of zinc chloride following the removal of cancerous tumors to avoid recurrence of the cancer.

> The cancer wards in the Middlesex Hospital are constantly forcing on the minds of the surgeons attached to it the questions, Can nothing be done to remove this fearful disease? Is, in the majority of cases, its recurrence after operation an inevitable necessity? May it not be that the apparent return of the disease is but the continued development of germs which were not included in the extirpation of the tumour? [4]

Two years earlier, a colleague had successfully applied solid chloride of zinc following extirpation of an extensive skin cancer of the face. He was ready to test the effectiveness of zinc chloride in a case of breast cancer with involvement of the glands of the armpit (axilla).

> Mr. Moore first detached all the axillary glands, and tying them above, cut them off. He immediately touched the stump with solid chloride of zinc. He then dissected out the diseased mammary gland, and before closing the wound he lightly touched nearly all the exposed axillary tissue with the chloride. The case ultimately did well, but what was remarkable was, while the skin over the breast sloughed to a slight extent, and was inflamed for some distance, due in great measure, according to Mr. Moore's notes, to its being somewhat tightly held together by the sutures, there "was no tension or great discharge from the axilla," and after a fortnight there is a report that the axilla is well filled up. It was clear that, at any rate, the touch with the caustic did not retard the cure.[5]

It occurred to De Morgan "that one might obtain the benefits that were sought by using the caustic in a less active form, and that a strong lotion of the chloride of zinc applied freely over the whole exposed area, after an operation for the removal of cancer, would penetrate to some little extent beyond the limits of the section, and would at least destroy any floating particles of the disease which might adhere to it without endangering the vitality of the whole thickness of the flap." [6] He employed solutions, increasing in strength, in a series of cases and, while he admitted the absurdity of forming conclusions on such limited data, he was himself satisfied with the results.

> The singularly favourable way in which the wound healed in the first case on which I tried this plan, satisfied me that it might be beneficially adopted in other than cancerous cases. There was one point which especially struck me, as giving it great value in hospital practice: the perfect purity of the discharges from the wound during the first few days after operation. It is well known that the presence of decomposing animal matter tends to bring any dead animal matter with which it may be in contact into a rapid state of decomposition; and if this takes place in a wound, it will certainly interfere for a time with natural and healthy processes of cure.[7]

(At about the same time Joseph Lister [1827–1912] was developing his concept of antisepsis. He first employed zinc chloride but later turned to carbolic acid.) Other surgical cases to which De Morgan applied his solution included amputations (even with extensive and thin flaps), operations in the area of the rectum involving mucous membrane and in the pelvic region, and accidental wounds. "In many cases the wounds have healed in twenty-four hours, without the least fulness or swelling, and leaving a [scar] line . . . which after a short time could hardly be seen or felt." [8] De Morgan concluded that he had "met with nothing which acts in so unmistakably beneficial a manner as the chloride of zinc." [9]

After a lapse of two years, by which time Lister had published his paper on the antiseptic principles in the practice of surgery,[10] De Morgan was thinking of his zinc chloride solution as an antiseptic. "It is not my object in this communication," he began a paper read to the Clinical Society of London on 8 May 1868, "to provoke a discussion of the comparative merits of the various antiseptics now in use, but rather to insist on their utility in general—as local applications—through illustrating the fact by reference to the action of chloride of zinc only." [11]

The blood, serum, and debris of tissue lying in a recent wound, he pointed out, will almost certainly become decomposed in a short time if left to themselves. Zinc chloride "forms with the albuminous elements a coagulum over the surface of the wound. This coagulum is incapable of decomposition, and it—and any fluids which may exude into the cavity of the wound—are kept free from taint." [12]

De Morgan questioned whether it is the presence of living germs in the atmosphere alone that causes putrefactive decomposition to occur. He conceded that the air is loaded with animal and vegetable germs that may set up a process of decomposition in animal fluid, but he observed that, when air is excluded from a wound or cavity, while decomposition may not occur as readily, it may still occur. This led him to speculate that germs in the tissues themselves may, under favorable conditions, take on active growth, and to the conclusion that when an antiseptic (by which he meant zinc chloride) is applied to the tissues or their secretions, either the inherent germs are destroyed, depriving the atmosphere of its power to produce decomposition, or else the surface or secretion is so altered in character as to destroy germs deposited from the atmosphere.

> The cases I am about to mention show that where the chloride of zinc has been freely used in the treatment of abscess under the most unfavourable circumstances—that is, where in scrofulous subjects they have been connected with diseased joints—their cure has been as rapid as would be the case in abscesses of the same extent in the most healthy persons, placed under the most favourable conditions.[13]

After reviewing the cases De Morgan concluded:

> Now, I believe that these cases have all gone on as well as they have done in consequence of the healthy state in which the parts have been maintained immediately and for some time after the operations, through the agency of the antiseptic employed. . . .
>
> I do not pretend that every case will do as well. I have seen no instances in which harm has been done by the lotion; but some cases have occurred in which the suppuration has gone on as profusely, and for as long a time, as if the abscesses had been simply opened and left to run their natural course.[14]

In a paper read before the Boston Society for Medical Observation on 7 June 1886, Arthur H. Nichols noted that when he was a medical student in Paris in 1864 he witnessed experiments conducted by Jacques-Gilles-Thomas Maisonneuve (1809–1897), the enterprising and versatile surgeon at the Hôtel-Dieu, in the treatment of carcinoma, scirrhus, and other recurrent fibrous tumors. His treatment involved caustic arrows (chloride of zinc, one part; wheaten flour, three parts; enough water to create a paste; rolled to form a thin round sheet from which arrows of any dimension could be cut) thrust into deep incisions made with a bistoury at the edges of the mass to be removed. In the course of twelve or fourteen days the diseased structure was destroyed and detached, leaving a clean healthy surface that was speedily healed by scar tissue. Six years of experimentation had confirmed the efficacy of this method. "I saw enough to convince me, however, that the pain produced by this process was immeasurably worse than that inflicted by the knife, and, of course, of much longer duration." [15]

The main drive of Nichols's paper was against deaths resulting from the (often accidental) ingestion of zinc chloride, with particular reference to a case in which he became involved. The case is so revealing of quackery ("irregular practice") at the close of the last century that Nichols's account is worth quoting at length:

> On the 7th of January, 1886, I received an urgent summons to visit J. R., residing in Dorchester. In the month of June previous, this man, whose family I had for many years attended, presented himself for advice regarding an incipient epithelioma [skin cancer] situated on the margin of the lower lip. The imperfect development of the growth seemed to render immediate surgical interference unnecessary, and delay was accordingly recommended. This advice did not prove satisfactory, and during the summer he put himself under the care of an irregular practitioner from Worcester, whose treatment proved wholly ineffective. Subsequently, attracted by a mendacious advertisement, and the reputed cure of a neighbor by another irregular physician from Lynn, the latter was called upon, and readily agreed to effect a prompt cure. To quote from this advertisement, now before me, the writer states that he has "a practice peculiarly his own, having never divulged it to any one." He furthermore alleges that "the medicine used is a vegetable liquid, and is a complete and effectual antidote for cancers and tumors, affecting only cancerous tissue. . . . In killing an ordinary cancer, he requires only from two to six hours, and the pain is so slight, a child can easily bear it."
>
> Upon reaching the patient about 7, P.M., I found him unconscious, respiration slow and stertorous; . . . face flushed, and whole body bathed in cold perspiration.
>
> It transpired that at 1, P.M., of that day, a son and partner of this physician, a young man of extremely limited professional at-

> tainments, had come to the house and persuaded the family that he was in all respects competent to conduct the treatment, having received the required instructions and remedial agents from his father. The patient's apprehensions were quieted with the assurance, reiterated in the presence of the family, that the medicines employed were purely vegetable, and that the pain produced would be insignificant. The lip was then treated with two different applications, the latter being left bound upon the diseased surface when the young man took his departure at the expiration of about an hour and a half. . . .
>
> From the very outset, the patient protested that the pain produced by these applications was unendurable, and it gradually increased in intensity, extending beneath the chin to the neck. These complaints were treated with ridicule by the attendant, who assured him that the drugs used might be even *swallowed* with impunity—a statement that may be important as tending to throw light upon the real cause of the accident which followed. At about 3, P.M., he complained of the region of the stomach, and soon after rushed to a rear door, seemingly to get fresh air. A violent chill now set in, associated with a numbness of the lower extremities, and a sensation of dizziness. He was then assisted to a lounge, and the plaster removed from his lip. Convulsive movements of the upper extremities now supervened, and he soon sank into a state of stupor, followed by a coma, from which he could never after be aroused.

The patient died at 9:30 P.M. Prior to that the "physician from Lynn," who had been sent for, admitted that, while the first application had been vegetable, the second, which had been "left bound upon the diseased surface," had been a caustic paste made of zinc chloride, starch, and mandrake.

"The patient . . . was a robust, rugged farmer, aged fifty-two, of steady, temperate habits; he had been in perfect physical condition up to the very day of his decease, with the exception of an occasional slight attack of neuralgic rheumatism." Part of the autopsy report read: "If we accept the statement of the attendant and family concerning the treatment and subsequent behavior of this man, it is highly probable that death was due to the absorption by the *primæ viæ* [alimentary canal] of a portion of the strong mineral caustic (chloride of zinc), which trickled into the mouth from the plaster applied to the margin of the lip, the ingested portion proving fatal by the profound shock exerted upon the nervous system." [16]

Nichols closed his discussion with reference to "sufferers, who, in consequence of some previous fright or personal idiosyncrasy, entertain a morbid dread of any operation involving hæmorrhage, and who prefer therefore to submit to [the] pain [of caustic treatment], rather than to the surgeon's knife. . . .

"It is, therefore, to be regretted, that so many patients are compelled, in order to obtain relief by this method of their choice, to resort to non-reputable practitioners, because the legitimate treatment by caustics formerly so highly commended has by reason of fashion, or from other insufficient causes, been substantially discarded by the regular profession." [17]

While the external and internal therapeutic use of zinc chloride rode the seesaw of popularity and condemnation, there was one area related to health in which there seems to have been no doubt about its efficacy. British navy medical reports for 1853 include one on "the effects of chloride of zinc in deodorizing offensive effluvia from cesspools, sewers, etc., and in decomposing poisonous emanations from the bodies of those affected by contagious diseases." [18] The report was well documented by navy surgeons and others and arrived at this conclusion:

> These extracts, taken from a large number of similar reports, are quite sufficient grounds for inducing us to recommend our readers to urge upon all with whom they have influence the benefit they may confer upon the community by having the chlo-

ride of zinc freely used in all the necessary operations for cleaning cesspools, privies, drains, etc., in the purification of unclean houses, and in all places where many people are crowded together, especially if any of them are suffering from the prevailing epidemic cholera.[19]

A quarter of a century later E. S. Horne and Company of Philadelphia were advertising their "Number One" chloride of zinc disinfectant "As Recommended by the Phila. Board of Health," the Board's medical inspector, J. Howard Taylor, stating:

> I have carefully tested the sample of Chloride of Zinc (your "Number One" Disinfectant,) left with me, and found it to be thoroughly satisfactory. Chloride of Zinc is highly antiseptic, and hence one of the best known substances for checking animal decomposition, and holding the gases resulting from putrefaction. It is convenient for household use, active in small quantities, and so long as maintained at its present standard, a thoroughly reliable article.

According to its promoters, the disinfectant destroyed the germs of diphtheria, typhoid, and yellow fever "that lurk in cesspools, and the imperfect underground drainage of the present day." It was further claimed that it would promptly do away with offensive smells in water closets, urinals, kitchen sinks, waste pipes, privy wells, foul cellars, and stables. "It is invaluable in the sick room, and to Undertakers, Surgeons and Physicians on Shipboard, in Prisons, Public Institutions, Factories, School Houses, Hospitals, . . ." To contain cooking odors in hotels, restaurants, and private homes, "saturate a large towel with 'NUMBER ONE,' diluted with water, and suspend on a line in your kitchen, near the inner door. . . . Do not use soap in washing the towel." [20]

Magnesium

There was a drought in England in the summer of 1618. Henry Wicker of Epsom, a large village about seventeen miles south of London, was in search of water for his cattle. Finding a small hole filled with water, he enlarged it so that his animals might drink. Thirsty as they were, they refused this offering. The water was bitter-tasting. Before long it was found that it not only healed external ulcers but was beneficial when taken internally.

For ten years only the local inhabitants made use of this medicinal spring. Then Lord Dudley North, "laboring under a melancholy disposition, drank the mineral waters as a medicine and found it to be an efficient purgative." [21] "Epsom water" was quickly accepted as an internal remedy and purifier of the blood. During the reign of Charles II Epsom became a fashionable playground of the rich—so much so that, in 1690, the Lord of the Manor, a Mr. Parkhurst, extended the existing buildings (started in 1620 by a predecessor in the office who erected a shed to shelter the visiting sick) to include a ballroom seventy feet long. In 1695 the botanical histologist Nehemiah Grew (1641–1712) isolated magnesium sulfate from the Epsom water.

By 1700 Epsom was averaging two thousand visitors daily, but by 1715, thanks to the manipulations of a scheming apothecary named John Livingston, the waters had fallen into disrepute. Meanwhile, in 1700, George and Francis Moult had begun the production and sale of magnesium sulfate from a spring at Shooter's Hill, in the vicinity of London. In 1717 Hoffmann found magnesium sulfate in the waters of the Seidlitz spring in Germany. (While Epsom salt is still magnesium sulfate, Seidlitz powder today does not include any form of magnesium in its composition.)

(In the summer of 1792, three gentlemen hunting in the

vicinity of Saratoga, New York, chanced upon a small stream of water issuing from an aperture in the side of a rock. The water was strongly mineral. Its source was given the name of Congress Spring. Since subsequent analysis revealed that the water contained about two-thirds chloride of sodium to one-sixth each of carbonate of lime and bicarbonate of magnesium, it could not be compared with the magnesium sulfate springs of Epsom and Seidlitz.) [22]

At the beginning of the eighteenth century the Count di Palma of Rome was marketing a white powder (magnesia alba), credited with almost unlimited medical virtues. He attempted to conceal the method of preparation, but in 1707 Michael Bernard Valentini (1657–1729) of Giessen, Germany, made public a process by which a similar powder could be obtained as a by-product of the preparation of nitre (potassium nitrate). Two years later Joannes Hadrianus Slevogt (1653–1726) produced a powder exactly resembling the Count di Palma's by precipitating magnesia from a solution of the sulfate through the addition of potassium carbonate (potash). In 1717 Giovanni Maria Lancisi (1655–1722) of Rome, the greatest clinician of his day, commented on the process. Five years later Hoffmann added potassium carbonate to magnesium chloride (the mother liquid left in the preparation of nitre) to precipitate magnesium carbonate (magnesia alba). (Magnesia alba was admitted to the London Pharmacopoeia in 1787.)

In 1754 the great Scottish chemist Joseph Black (1728–1799) established the characteristics of magnesia and its salts, but he shared the then popular opinion that magnesia, actually the oxide of magnesium, was the true metal, a misconception that persisted until 1808, when the metal was isolated by the celebrated English chemist Humphry Davy (1788–1829).

Black had begun an investigation of limewater as a possible

agent for dissolving calculi while at the University of Glasgow. (He transferred to Edinburgh in 1752.) It has been suggested that he passed directly from limewater to an investigation of magnesia alba, but this is almost certainly not the case for on 10 February 1753, he wrote Professor William Cullen (1712–1790), his respected tutor and mentor at Glasgow, that, for lack of time, he had not read a single work on limewater and had, in fact, "almost forgotten all my Glasgow projects." [23] More likely, Black turned to magnesia alba early in 1754. In four and a half months he produced a dissertation on the subject that "the medical faculty of Edinburgh could see . . . was more than worthy of their approbation." [24]

Why magnesia alba? A popular view has been that the "choice was simply a logical extension of the limewater project instead of a 'laying aside' of the latter as Black clearly saw it *at the time.*" But internal evidence supports an opposite view. "Certainly none of the few simple experiments, which were all Black would have made to see if the magnesia alba would 'yield a limewater that might be more effective than the common sort' (it did not), would have led into the numerous experiments reported in the dissertation, or even into a reasonable dissertation project per se." There were obviously additional factors. "Magnesia alba as a part of contemporary materia medica had a history which, if less politically explosive than that of limewater, had an intrinsic interest vis-à-vis the materia medica. First investigated scientifically by the great German physician and chemist Friedrich Hoffmann in the later seventeenth century, both the therapeutic action of magnesia alba and the chemically correct method of its synthesis were still matters of debate in British pharmacopoeias midway through the eighteenth century." [25] And by Black's own account, "I was indeed led to this examination of the absorbent earths, partly by the hope of discovering a new sort of lime and lime-water, which might possibly be a more powerful solvent of the stone than that commonly used; but was disappointed in my expectations." [26]

Humphry Davy is popularly remembered as the developer of the Davy safety lamp. This oil-burning lamp was covered with wire gauze which absorbed the heat of the flame so that it would not cause an explosion when brought into the vicinity of inflammable gas. Davy took out no patent on the lamp, which was extensively used in coal mines, and in 1817—one year after its invention—the grateful coalminers of Newcastle presented Davy with a silver-plate dinner service. He received a baronetcy in 1818 for his service to industry.

The lamp was a by-product of Davy's major investigations, which were in the field of electrochemistry and gasses. At the beginning of the nineteenth century it was believed that sodium hydroxide and potassium hydroxide were elements. In 1807 Davy succeeded in isolating sodium and potassium from these compounds by electrolysis. His success prompted him to attempt to decompose the alkaline and common earths, which included magnesia. (Magnesium has since been classified as a common metal.) Other earths with which Davy experimented were barytes (barium oxide), strontites (oxide of strontium), and lime (calcium oxide). His results were highly unsatisfactory prior to June 1808, when he received a letter from the Swedish chemist Jöns Jakob Berzelius (1779–1848) who in 1829 would discover thorium. Davy learned that Berzelius "had succeeded in decomposing barytes and lime, by negatively electrifying mercury in contact with them, and that in this way he had obtained amalgams of the metals of these earths.

"I immediately repeated these operations with perfect success. . . .

"That the same happy methods must succeed with strontites and magnesia, it was not easy to doubt, and I quickly tried the experiment.

"From strontites I obtained a very rapid result; but from magnesia, in the first trials, no amalgam could be procured. By continuing the process, however, for a longer time, and keeping the earth continually moist, at last a combination of the basis with mercury was obtained, which slowly produced magnesia by absorption of oxygene from air, or by action of water. . . .

"I was inclined to believe that one reason why magnesia was less easy to metallize than the other alkaline earths, was its insoluability in water. . . . On this idea I tried the experiment, using moistened sulphate of magnesia, instead of the pure earth; and I found that the amalgam was much sooner obtained." [27]

Davy found that complete decomposition was very difficult because to accomplish it "nearly a red heat was required, and at a red heat the bases of the earths instantly acted upon the glass [tubes used in the process], and became oxygenated." There were other difficulties, as a consequence of which "in a multitude of trials, I obtained only a few successful results, and in no case could I be absolutely certain that there was not a minute portion of mercury still in combination with the metals of the earths." [28]

> The metal from magnesia seemed to act upon the glass, even before the whole of the quicksilver was distilled from it. In an experiment in which I stopped the process before the mercury was entirely driven off, it appeared as a solid, having the same whiteness and lustre as the other metals of the earths. It sunk rapidly in water, though surrounded by globules of gas, producing magnesia, and quickly changed in air, becoming covered with a white crust, and falling into a fine powder, which proved to be magnesia.[29]

All four of Davy's newly isolated metals were highly unstable, but they were nonetheless deserving of names, "and on the same principle as I have named the bases of the fixed alkalies, potassium and sodium, I shall venture to denominate the metals from the alkaline earths barium, strontium, calcium, and magnium; the last of these words is undoubtedly objectionable, but magnesium has already been applied to metallic manganese, and would consequently have been an equivocal term." [30] (*Magnium* has, however, long given place to *magnesium*.)

Notwithstanding the claim of the Count di Palma that his powder was a panacea, the medicinal use of magnesium salts was largely as a purgative—at best a cleanser or purifier of the body.

> Every physician will of course be guided in the choice of his purgatives by the peculiar circumstances of each case, but there is no purgative so generally applicable to all cases, so safe, so agreeable, and at the same time so efficacious, as the acid saline solution [sulphate of magnesia plus the "*dilute* sulphuric acid of the Dublin or Edinburgh pharmacopœias"]. It is the reproach of our art that means which we employ to remove disease are almost always disagreeable, sometimes as disagreeable as the diseases themselves.
>
> Our purgatives particularly subject us to this reproach, and in an especial degree, our liquid purgatives. If I shall have succeeded in introducing to general notice a liquid purgative which is not disagreeable, either to taste or smell or sight, while at the same time it is efficacious, without producing sickness or griping, I shall feel that I have contributed somewhat to lessen the reproach hitherto but justly cast upon our art, and that the time I have bestowed on this subject has not been thrown away. [31]

In 1905 and 1906 Latvian-born physiologist Samuel James Meltzer (1851–1920) and his associate John Auer (1875–1948) at the Rockefeller Institute investigated the action of the magnesium salts on tetanus. They also demonstrated the anesthetic properties of magnesium. In 1913 the noted anesthesiologist James Taylor Gwathmey (1863–1944) proposed rectal injection of ether and oil preceded by morphine and magnesium sulphate injections to replace inhalation anesthesia; four years later, Gwathmey and Howard T. Karsner, celebrated Western Reserve

(Cleveland) pathologist, proposed oral administration of this mixture to deaden pain in the changing of dressings. By 1922 Gwathmey had improved his method of *synergistic anesthesia* (in which the total effect of two drugs working together exceeds the sum of the two effects acting independently) and a year later, in obstetrical cases, was using intramuscular injections of morphine and magnesium sulfate combined with rectal injection of quinine, alcohol, and ether in olive oil. In 1925 Edmond Meyer Lazard introduced intravenous injections of magnesium sulfate in eclampsia, a toxic disturbance in pregnancy involving convulsions, a practice still followed. Magnesium sulfate is also given (intramuscularly or intravenously) to control uterine tetany, a nervous affection characterized by spasms.

Tin

Tin came into medical use in the Middle Ages largely as an agent for expelling intestinal worms, a purpose for which it was recommended by Paracelsus. The alchemists produced mosaic gold (*aurum musivum*) by combining tin and mercury into an amalgam and then distilling this substance with sulfur and chloride of ammonia. Mosaic gold, later to be known as bisulfide of tin, was a golden solid of crystalline structure whose brilliant luster appealed to the eye of the alchemist. This was the first tin compound to be used in medicine. In addition to serving as a vermifuge (worm expellant), it was employed as a sweat inducer and purgative in fevers, hysterical complaints, and venereal disorders. Subsequently, the binoxide, the nitrate, and sometimes the chloride (discovered by Libavius at the close of the sixteenth century) of tin were used as vermifuges, but they were ultimately replaced by powdered tin given either with chalk, sugar, and crabs' eyes, or combined with honey or some sweetmeat. Antihecticum Poterii (a combination of tin with iron and antimony to which nitrate of potash was added) was thought to be especially useful as a sudorific (sweat inducer) in cases of consumption and blood spitting. An antihæmorrhoidal ointment known as Flake's was an amalgam of tin combined with rose ointment and red mercuric oxide.

Compounds of tin as medicines were popular with the regular physicians, the quacks, and the promoters of patent medicines through the seventeenth and eighteenth centuries, but thereafter fell into disuse. Early in the nineteenth century, however, some doctors turned to the hydrochlorate of tin as the most effective substitute for alcohol (which they regarded as too highly priced) in the preservation of anatomical specimens. By the 1860s Lister was employing block tin (or tinfoil strengthened with adhesive plaster) in surgical and abscess cases as a covering to ensure the efficiency of the fomentations (by preventing evaporation of carbolic acid) and to serve as a splint.

Pewter is an alloy of which the essential ingredient is tin, with lead added in varying proportions. The amount of lead in "good pewter" may not exceed 20 percent, although a dark soft pewter, known as ley, contains up to 50 percent. The addition of a small amount of copper produces "hard" pewter, while the substitution of antimony for copper lends the alloy a silvery luster.

Pewter's relationship to medicine lies in various utensils made of the alloy. Pewter medicine spoons and syringes date back to around 1700. From as early as 1565 pap (flour or bread cooked in water with milk occasionally added) was employed for the supplementary, sometimes the sole, feeding of infants, and by the eighteenth century pewter pap boats, with a capacity of 2½ to 3 fluid ounces, were being used for such feedings. In the same century the English, French, and Dutch employed pewter infant feeding flasks (the Chinese followed a century later) with a capacity of 6 to 10 ounces. These flasks, which unscrewed at the neck for cleaning, were provided with spongy nipples made of cloth or leather, suitably perforated. The eighteenth century also saw pewter bedpans, male urinals, chamber pots, and commode liners, pewter being preferred as a material for these utensils because of its potential durability. In the eighteenth and nineteenth centuries there were pewter croup kettles or inhalers, bleeding bowls, and spouted feeding pots.

Iron

In 1574 Nicholas Monardes (1493–1588), a physician of Seville, Spain, published his *Historia Medicinal*. While a dedicated student of the flora of the New World, Monardes was not irrevocably committed to vegetable drugs. The last part of the second volume of his history is a dialogue on iron and its virtues.

But despite this, and occasional passing references to its earlier use, iron was not considered of special medicinal value until the seventeenth and eighteenth centuries.

It has already been noted that the seventeenth-century chemist Nicholas Lémery discovered iron in the ashes of animal tissues. About a century later V. Menghini (fl. 1745) showed that this iron was not in the flesh or bones but in the blood, and one portion of the blood only—the corpuscles. (Berzelius subsequently verified Menghini's findings.)

While these discoveries gave impetus to a burgeoning ferrous therapy, it is to Thomas Sydenham and his contemporary, the anatomist and exponent of chemical medicine Thomas Willis (1621–1675), that the general adoption of iron as a medicine must be credited.

Willis employed a secret preparation of iron, of which Walter Harris (1647–1732), a famous follower and protégé of Sydenham and physician-in-ordinary to Charles II, wrote in his *Pharmacologia Anti-Empirica* (1683): "The best preparation of any that iron can yield us is a secret of Dr. Willis. It has hitherto been a great secret and sold at a great price. It was known as Dr. Willis's Preparation of Steel." He then proceeded to reveal that Willis's formula was equal parts of iron filings and crude tartar powdered and mixed with water in a glazed earthen vessel. The resultant damp mass was dried over a slow fire or in the sun. It was then wetted and dried again, a process repeated four or five times. It was given in white wine, as a syrup, or made into pills, an electuary (sweetmeat), or lozenges. Its major function was the removal of obstructions in the bloodstream. John Quincy, who published his English medical dictionary in 1719, in describing the action of iron in removing such obstructions, said that, the pull of gravity being greater on a metallic particle than on lighter particles in the bloodstream, the former did a better job.[32]

Sydenham, in a dissertation published in 1682, traced hysteric diseases to the animal spirits not being rightly disposed and indicated that treatment must be directed to strengthening the blood—the fountain and origin of the spirits. He recommended in all diseases involving anemia, bleeding (if the patient was strong enough) followed by a course of ferrous therapy.

> Then he described, much the same as modern treatises do, how rapidly iron quickens the pulses, and freshens the pale countenances. In his experience he had found that it is better to give it in substance than in any of the preparations, "for busy chemists make this as well as other excellent medicines worse rather than better by their perverse and over officious diligence." He advises 8 grains of steel filings made into two pills with extract of wormwood to be taken early in the morning and at 5 p.m. for thirty days; a draught of wormwood wine to follow each dose. "Next to the steel in substance," he adds, "I choose the syrup of it prepared with filings of steel or iron infused in cold Rhenish wine till the wine is sufficiently impregnated, and afterwards strained and boiled to the consistence of a syrup with a sufficient quantity of sugar."[33]

(A century and a half later Berzelius would show that iron is closely associated with the coloring matter of the blood or, as it would come to be called, the hemoglobin.)

Moses Griffith (1724–1785), son of a Welsh tax collector, received his M.D. degree from Leyden on 30 October 1744. After practicing in London for better than twenty years, he moved to Colchester in Essex (about 50 miles northeast of the capital) in 1768 and eight years later published his *Practical Observations on the Cure of Hectic and Slow Fevers, and the Pulmonary Consumption.* He had found a mixture involving ferrous sulfate, potassium carbonate, myrrh, and sugar to be effective in treating

> slow fever attended with a slow pulse, not much heat, loss of appetite, dejection of spirits, restless nights, and disturbed sleeps. . . . Likewise after long and severe fevers, that have broken down the constitution and are often succeeded by lowness, want of appetite, and night sweat. . . . In the slow fever, which often attends a *chlorosis* [iron deficiency anemia]. . . . Also, after the blood has been drained, and the body weakened, by a large discharge of matter from a succession of abscesses. . . . After large hemorrhages, which leave behind them a slow fever.

The feature common to these maladies seems to be that recognized by Sydenham—iron deficiency anemia. Griffith's compound iron mixture could be expected to improve the condition of such patients.[34]

Griffith's mixture, of which the active ingredient was ferrous carbonate, found its way into the 1809 edition of the *London Pharmacopoeia,* which also listed a solid form—Griffith's pill. While ferrous carbonate came to be regarded as the most easily absorbed of the iron preparations, it offered the disadvantage of being unstable. In the preface to his book, Griffith anticipated this objection and attempted to brush it aside by pointing out that one cannot "determine the effects of a medicine, when taken into the body; where in mixing with the juices of the stomach and intestines, it may undergo an alteration, which no analysis out of the body, can ascertain." [35] In later days the instability factor was overcome by requiring that the mixture or pill be made fresh for each patient.

Tartrate of iron, long a favorite form for the administration of iron, was admitted to the pharmacopoeia in the eighteenth century. A complicated formula, *boules de Mars,* continued into the twentieth century as a popular French remedy.

In 1882 the anonymous author of *A Study in the Action of Iron* pointed to two theories that had been advanced to explain the good effects following the administration of iron: "the first, that [they] are produced by its action on the red blood corpuscle; the second, that it exerts a special influence on the digestive system." [36]

The argument in support of the theories went like this: Iron is a constituent of the red corpuscle; in anemia the number of red corpuscles is below normal; the introduction of iron increases the number of red corpuscles. But the total amount of iron in the human system does not exceed 3 grams (about 45 grains), almost all of which is in the red corpuscles. Only a portion of this blood-iron will be lost at any one time and, since there is some iron in practically all food, the deficiency can never be very great. Furthermore, it is not through a few doses of iron, but only after continual use, that good effects are obtained and, as Roberts Bartholow (1831–1904), one of the founders of the American Neurological Association, has pointed out, "iron improves but little if at all the conditions of the anemic when it does not increase the desire for food and the ability to digest it." According to the great French physiologist Claude Bernard (1813–1878), the salts of iron have a special action on the mucous membrane of the stomach and all parts with which it comes in contact take on a more active circulation.[37]

Many of the nineteenth-century accounts of the medical use of iron favor one theory or the other though they sometimes embrace both. In 1839 T. C. Adam of Lenawee County, Michigan, reported that

> for upwards of five years we have been in the habit of prescribing, almost daily, the *liquor ferri persesquinitratis,* . . . [and] have derived from its use very remarkable assistance in the treatment of several diseases, especially diarrhœa, and other affections of mucous membranes accompanied by discharges; . . . In cases . . . of habitual diarrhœa, from birth perhaps, in children, and in cases in which there seems to exist an excess or irritability in the digestive tube—we know of no medicine which produces a more beneficial, immediate effect; and certainly, in its power of preventing similar attacks in future, this remedy is without any rival, so far as our experience extends. In cases of children, we have found its long-continued employment produce the most satisfactory results.[38]

Twenty years later, however, Isaac Casselberry of Evansville, Indiana, had this to say:

> *Fever is a diseased transformation of all the tissues;* but the fluid tissues suffer the most, because the solid tissues are formed in them and of them. . . .
>
> After . . . the removal of the most manifest symptoms of fever, there often remains an impoverished condition of the blood . . . [which] is *sensitively* evinced by lesion of nutrition, loss of nervous energy, want of appetite, muscular debility, and more or less perversion of secretion; and *it is caused by deficient development intensity of the molecular combinations of the elements of the blood.* . . .
>
> The physiological effects of iron conclusively evince, that it promotes the development of the blood-cells and accelerates their maturity. This is in accordance with a general law of human or-

ganism, that the specific stimulants of cell growth in every tissue are elements identical with the natural contents of the cells, or convertible into them.

The globules of the blood contain certain increments of iron obtained from the food; and from the physiological fact that these are always present in normal blood, it is self-evident that iron is absolutely necessary to animal life. . . .

Iron is the most efficient agent to promote . . . normal restoration . . . because it supplies the element required to promote the growth and maturity of protein globules. . . .[39]

An epidemic of diphtheria in 1859 enabled R. W. Crighton of Leamington, in the English Midlands, to make an extensive test of the effects of iron on this disease. This and other experimentation led Crighton to the conclusion that muriated tincture of iron (combined with ammonium acetate) was invaluable in all inflammatory disorders.

In 1862 E. N. Chapman of the Long Island College Hospital, Brooklyn, New York, was extolling the virtues of the pyrophosphate of iron:

The citro-ammoniacal pyrophosphate of iron affords certain marked advantages over the preparations of iron hitherto in use. Its tastelessness, in solution with sugar, and elegant appearance, in our day, when the nauseous doses of the older practitioners will not be tolerated, is an important item in the care of children, or adults even, when the employment of a remedy is demanded for a period of time. There is every reason for prescribing our medicines in as palatable and pleasant a form as possible. In addition, there are many persons of a nervous, delicate organization, particularly females, who cannot take the ordinary preparations of iron. They disorder the stomach—in their language are too heating—and thus not only fail to be assimilated, but, by perverting the gastric and intestinal secretions, seriously interfere with the digestion. Hence, instead of enriching the blood by new materials, we are merely cutting off the original supply, imperfect as it is, and making the gastric surface a centre of morbid irritation. . . .

A marked peculiarity in the pyrophosphate of iron is the fact that it will, scarcely ever, . . . disagree, and very frequently patients who cannot tolerate the ordinary forms of iron, will bear this well, and receive great benefit from its use. . . . The new salt will supply the iron to the blood-globules as promptly, but not more so, than the others.

. . . All the common preparations of iron are apt to oppress the stomach, coat the tongue and destroy the appetite, especially when the patient is much debilitated. . . . The pyrophosphate is friendly to the stomach, will never cause any irritation of the gastric surfaces, and, to our knowledge, has never disagreed with any patient, however incompatible the other forms may have been. Besides, it appears to possess a tonic power, and will restore the appetite and digestion after the failure of bitters, quinine, wine, &c.[40]

In 1876 John C. Lucas, a surgeon in Her Majesty's Indian Army, reported that all preparations of iron, but especially the perchloride and the pernitrate, were valuable as therapeutic, antiseptic, and preventive agents in such maladies as enteric fever, cholera, septicemia, erysipelas, debility, and puerperal fever. Three years later T. Grainger Stewart of the University of Edinburgh was recommending large doses of iron in certain cardiac cases, particularly those involving diseased aortic valves. At about the same time West Indian-born Jacob Mendes Da Costa (1833–1900) of Philadelphia, perhaps the ablest clinical teacher of his time in the eastern states, introduced the hypodermic injection of dialyzed iron in severe cases of anemia.

Iron was used to cope with some of the problems of women. Both Adam and Casselberry employed it in one form or another for excessive bleeding during the menstrual period, and

Adam used nitrate of iron in other cases of abnormal vaginal discharge.

Around 1850, in the case of a woman who had lost three babies prematurely, notwithstanding the fact that "she had two sisters married, whom she considered much more weakly and diseased than herself, and yet they had living and ordinary healthy children," Abraham Livezey of Lumberville, Pennsylvania, decided to use "iron scales, as they fall from the smith's anvil, steeped in hard cider, from the beginning to the end of the next pregnancy. . . . [The patient's] appetite increased, her digestion, health and spirits improved, and a new hope dawned upon her. She took gallons of cider, rendered turbid, or somewhat inky, by the iron scales, and at the full term I was present at the delivery of a firm, full-grown male child. He grew finely, and waxed so strong by the ninth month that he could walk; and now, in his fifth year, he is remarkable for his tallness and strength—having born the cognomen from birth of the 'iron baby.' "[41]

But the therapeutic use of iron was not without its detractors. In 1877 a London chest physician, J. Milner Fothergill, published an article, "When Not to Give Iron," in which, while admitting that "conditions which call for the administration of iron are *par excellence* those where debility is combined with anæmia," he stressed that the use of iron in certain other circumstances in which it had been employed by physicians was contraindicated. The following year R. W. Crighton, who had by now moved on to the Tavistock Dispensary near Plymouth, while taking issue in two cases rejected by Fothergill, found that his observations coincided "in most instances with my own experience."[42] In 1879 the English gynecologist J. Matthews Duncan reported negatively on the use of perchloride of iron as a styptic or uterine irritant in cases of postpartum hemorrhage.

Modern medical science recognizes that iron is essential to

the normal transportation of oxygen in the body and to normal tissue respiration. The need for iron replacement is particularly vital in women.

Lead

The great eighteenth-century authority on the medicinal use of lead was Thomas Goulard (d. ca. 1784), a surgeon of Montpellier (France) who had "rather more than a local reputation. He was counsellor to the king, perpetual mayor of the town of Alet, lecturer and demonstrator royal in surgery, demonstrator royal of anatomy at the College of Physicians, fellow of the Royal Academies of Sciences in Montpellier, Toulouse, Lyons, and Nancy, pensioner of the king and of the province of Languedoc for lithotomy, and surgeon to the Military Hospital of Montpellier." [43]

His treatise on various preparations of lead, particularly the Extract of Saturn (lead was associated with Saturn from classical times), was published about the middle of the century:

> It is to chance alone we are indebted for the major part of the remedies now successfully made use of in surgery and physic. Nature often conceals from the learned those secrets, which she afterwards pleases to communicate to the ignorant or less attentive. The discovery of the peculiar properties of bark and mercury was more the effect of chance than erudition. But though we are indebted to the former for the most approved remedies, it is not from that alone we can ever learn the proper application of them: this must be the work of time, or discernment, and experience. . . . Amongst the rest, no one, I think, deserves a higher rank than Lead. Nay, I have reason to flatter myself, that when the reader has perused the following Treatise, he will be inclined to think this metal one of the most efficacious remedies for the cure of most diseases which require the assistance of surgery.
>
> It is true that, for many years past, its general virtues, as well

> as preparations, have been known; even some dispensatories partly describe the Extract of Saturn; but none give any account of the various modifications I made it undergo, and to which alone I attribute its surprising success.
>
> I by no means pretend to publish this composition of the Extract as any new discovery of my own: my only intent, in this Treatise, is to lay before the masters of the art, and the Public, the various forms (if I may so speak) I have given this Extract, and the particular cases in which it has been specific, to the no small astonishment of the Faculty.[44]

Extract of Saturn was prepared as follows: "Take as many pounds of litharge of gold [lead monoxide] as quarts of wine vinegar (if made of French wine the better); put them together into a glazed earthen pipkin, and let them boil, or rather simmer for an hour, or an hour and a quarter, taking care to stir them during the ebullition with a wooden spatula; take the vessell off the fire, let the whole settle, and then pour off the liquor which is upon the top into bottles for use." Water of Saturn, or Vegito-mineral Water, "is made by putting two tea-spoonfuls, or one hundred drops of the Extract of Saturn to a quart of water, and four tea-spoonfuls of brandy: the quantity of the Extract and brandy may be diminished or increased according to the nature of the disorder, or the greater or less degree of sensibility of the part grieved." Cerate of Saturn resulted from working four ounces of Extract of Saturn in six pounds of water into four ounces of refined wax and a pound of oil. To arrive at a Cataplasm (or Poultice): "Take some Vegeto-mineral Water, put it into a pipkin with crumbs of bread; let it just boil; lay a sufficient quantity of it on linen, and apply it to the part affected: this must be renewed three or four times in four-and-twenty hours . . ." Pomatum of Saturn was prepared by adding eighteen ounces of oil of roses, four or five ounces of Extract of Saturn, and a dram of camphor to eight ounces of melted refined wax. Nutritum of Saturn (effective against burns, skin diseases, and "disagreeable itchings") involved six ounces of litharge of gold reduced to a very fine powder and mixed in a mortar with five ounces of oil and eight ounces of Vegeto-mineral Water. Goulard continued with formulas for half-a-dozen pomatums and "plaisters." [45]

As for the therapeutic uses of his preparations: "An inflammation of the external parts of the body is a very common disease, and may proceed from a thousand different causes; such as a blow, a fall, a contusion, &c. from external accident: internally from any local defect of the solids; from the contraction of the smaller vessels; from their spasm; from the compression, constriction, and obstruction; from the difficulty which the blood finds in passing the extremity of the arteries; from its too great quantity, thickness, viscidity, mixture with heterogeneous and virulent parts; or from any other alteration, whether simple, or combined, &c." Goulard maintained that no medicine better subdues external inflammations than Extract of Saturn, since

> it is endowed with a singular property of penetrating the obstructed blood and lymphatic vessels, and of dispersing the inspissated matter therein, without too much relaxing or irritating the coats of the inflamed parts; it preserves a medium between these two actions, and thereby insensibly produces, without any bad consequences, the most surprising effects. This remedy seems to unite, at once, three qualities very essential for an antiphlogistic medicine; a cooling virtue, which the most ardent inflammatory heat cannot resist; an anodyne one, which quiets the most violent pains observed in inflammation; an attenuating, resolving quality, which the prejudiced part of mankind has unfairly confounded with repulsion: in short, all the parts of our body, without distinction, fatty, glandulous, muscular, tendinous, aponeurotic, membranous, ligamentinous, weak or strong in their texture, endowed with a greater or less degree of sensibility, bear equally the action of our metallic remedy.

For contusions,

> there is no application whose effect is so sure and expeditious as the Vegeto-mineral Water, warmed and applied by compresses to the part affected; taking care to moisten them from time to time. When the parts are lacerated, the wound is to be dressed with the Cerate of Saturn.
>
> Our method of cure [of burns] is very simple; it consists in the application of compresses dipped in the Vegeto-mineral Water, on the part which is burned. If the outward teguments remain unbroken, we need only keep the compresses constantly moistened with the above water; if the burn is deeper, if the teguments are injured, and there is a wound, little pieces of fine lint may be applied, covered with the Ointment [Nutritum of Saturn], . . . taking care to cover the whole with compresses dipped in the Vegeto-mineral Water, which must be moistened from time to time. By this means of proceeding, the pains are not only immediately calmed, but the disease radically cured.

Goulard assured practitioners who used Extract of Saturn in treating gunshot wounds that they would be spared

> the errors which attend the other methods hitherto adopted on the like occasions; for I am not afraid to assert, that in a short time after the application of this medicine, the accidents usually attending these complaints, such as inflammation, swelling, &c. will be diminished. I am also thoroughly persuaded, that by a proper attention in the application of our medicine, the cure of many gun-shot wounds might be effected, which are ranked among those which require amputation. . . . What a considerable loss do his Majesty's troops suffer, by the non-application of this Extract, both in the field and hospitals, where the wounded swarm! We have neglected nothing to make it public; and the most favourable experiments have justified our opinion: yet by an inconceivable infatuation, its use is by no means so general as it

> ought to be. Why have not modern authors, who have written *ex professo* on gun-shot wounds, and the manner of treating them, had candour enough to acknowledge the insufficiency, or rather the danger, of the most generally received processes? If they would only take the trouble to make trials of our application, the good effects of which we cannot sufficiently extol, perhaps they would be forced to declare in its favour.

Turning to external suppurations, a term Goulard employed to cover abscesses and ulcers, he said there is

> nothing more easy than the dressing of simple ulcers; I mean such as are occasioned by the opening of an abscess; for these we make use of washes of the Vegeto-mineral Water, and of injections of the same, when they are deep; we wet the pledgets and first compress; and cover the former with the Cerate. . . . It is not the same with old, foul, bleeding, horney, and fetid ulcers. . . . Having learned from experience that the Extract of Saturn, mixed with common water and brandy, . . . is not only sovereign in preventing putrefaction, cleansing, resolving, calming; but has the most singular virtue of blunting the acrimony of the most bloody and corrosive suppurations; I cannot think any other remedy can with equal success be applied to the cure of old ulcers. I dress with the ointment already mentioned, taking care to wash the ulcer with Vegeto-mineral Water a little warm, with which the pledgets and compresses are to be wetted, and to moisten every hour the bandages with the said water. . . .
>
> It is a received opinion among Physicians and Surgeons, that external applications are useless, and generally dangerous, in the treatment of cancerous tumours. . . . This opinion, so universally adopted, is the cause of these unhappy sufferers being left a sacrifice to the most horrible pains, which no medicine can calm, and which frequently conduct them to their graves. From hence one may judge of what importance it would be to discover a medicine, capable not only of relieving, but even of curing this otherwise invincible and cruel disorder. Now I have the comfort to

> think, that I have discovered such a medicine in the Extract of Saturn, . . . [applied] by way of lotion or cataplasm; because so far from shutting up the pores of the skin, it helps to open them; very far from heating the diseased part, it cools it; from exasperating and irritating, it helps to quiet. Again, by getting to the bottom of the grievance, it gives motion to the inspissated and stagnated fluids, without any inconvenience resulting from it; and if the application is continued, the cancerous tumour is observed to disperse, or at least it gives ease to the patient, which is no small point gained.

Sprains, stiffness of the joints, and relaxation of the ligaments were relieved by applications of Pomatum and Vegeto-mineral Water.

> A little attention to the manner in which the Extract of Saturn operates in rheumatic cases, is sufficient to convince us, that it possesses an attenuating, relaxing, and anodyne quality; so that the solution of the viscous, tough fluid, which occasioned the disease, far from producing the fatal accidents that so frequently attend the use of dispersers, and still more of repellers, operates by degrees, and without danger; first a diminution, and then a total cessation of the Rheumatic Pains, and of the other symptoms that accompanied them. When these pains are extremely obstinate, I apply the Saturnine plaister upon the part affected. . . . Extract of Saturn is also efficacious in periodical pains that attack the joints, and are termed gouty.

In the treatment of tetters (cutaneous diseases such as herpes, ringworm, and eczema), the

> metallic particles of our application insinuate themselves into the most minute pores of the skin, and penetrate to the only source which produced the tetterous complaint; when arrived there, they thin and divide the humour, blunt its acrimony,

> favour its egress through the pores of the skin, and at last radically cure the disease, sooner or later, according as the morbific matter opposes more or less resistance to the effect of our remedy. . . .
>
> The Itch is an eruption of little cutaneous pustules, which may be indiscriminately spread over the whole body, except the face; but is more particularly remarkable on the wrists, between the fingers, on the arms, hams, and thighs. . . . Itch . . . may be moist or dry. . . . Few diseases have had more remedies prescribed for them. . . . On the other hand, not only [are the effects of what I recommend] certain, but it is so inoffensive, that no exception can possibly be taken to the use of it. To these considerations I must beg leave to add another, which is suggested by the nature of his Majesty's service. It is certain that the military hospitals will be maintained at less expense, if my remedy be substituted in lieu of those which have hitherto been mentioned for the cure of the Itch.

Based on cures reported by hospitals at Arras, Bethune, and Gravelines,

> and those in the military one at Montpellier, wherein more than 2000 soldiers have made use of the Extract of Saturn, [it appears] that this medicine ought to be considered as the true specific for the treatment of the Itch. . . . Nevertheless, to hasten the scaling of the itchy eruptions, a little sea-water, and allum in powder, may be added to the Vegeto-mineral Water. . . . [To treat piles,] I wash the Piles with Vegeto-mineral Water, and then apply to them the Cerate, made with wax, oil, and the above water.[46]

The 1775 English edition of the Treatise ends with an announcement that "Mr. Goulard's Extract of Lead, made by himself, May be had at Mrs. Turmeau's, opposite Wedgewood and Bentley's Warehouse, in Greek-Street, Soho."

Goulard apparently gave no thought to the internal administration of lead, but before his time Adrian Mynsicht, physician to the Duke of Mecklenburg (Germany), who flourished early in the seventeenth century and is credited with having been the first to describe emetic tartar (1630), had devised a Powder of Saturn for the treatment of phthisis (pulmonary consumption) and asthma. The principal ingredient was Magistery of Saturn (white lead precipitated from the acetate by carbonate of potash). Other components were magistery of sulfur, squine root, flowers of sulfur, pearls, coral, oatmeal, Armenian bole, flowers of benzoin, olibanum, sugar candy, saffron, and cassia.

In 1806 Thomas R. Spence, himself an epileptic, proposed the use of lead acetate (sugar of lead) in the treatment of epilepsy. Twenty years later, William Laidlaw, an English surgeon, wrote:

> Very opposite opinions are entertained by individuals, equally distinguished for their candor and erudition, relative to the internal administration of the acetate of lead; some considering such use of it as a display of unqualified rashness, while others conceive it to be an invaluable remedy in many dangerous diseases; and that its prudent exhibition is not only justifiable, but highly laudable.
>
> That the incautious use of the acetate of lead will produce deleterious effects, would be equally idle and uncandid to deny. . . . But the production of so baneful a complaint [as colic], by the gradual introduction of this salt into the system, is not a conclusive proof, or rather is no proof at all, that its judicious use may not effect salutary results.
>
> Were this not admitted, then arsenic, the oxymuriate of mercury, and many other articles of the materia medica, with equal, nay, with greater propriety, might be discarded.[47]

The erratic climate of Georgia and Alabama tended to produce fevers that were more violent and unyielding than those encountered elsewhere. Reflecting "on the influence of certain cooling and sedative remedies in external injuries, such as fractures, sprains, &c., where great fever and irritability were developed," a Dr. A. Kimbal (of Macon County, Alabama) came to the realization that

> in such cases no local application had been of greater efficacy than the solution of acetate of lead. When externally applied, if its effects be to ease pain, by reducing inflammation and quieting the irritability of the parts, then if internally used, so that its application, to the inflamed tissue should be general, would it not be a happy and salutary adjuvant in the treatment of fevers? Acting upon this suggestion, I determined to give it a fair trial. Governed by the generally received opinion of the dangerous consequences of long retention of the article in the stomach and bowels; I was induced to unite with it such purgatives as would insure its speedy transition through the alimentary canal.

After reviewing a number of cases in which he successfully employed acetate of lead, Kimbal reached these conclusions:

> I cannot suppose that the beneficial effects of the acetate arise from its purgative property. It acts in three different modes. 1st. As an astringent, by placing a proper restrictive influence over the capillaries, and coordinating the inordinate alvin secretion. 2d. As a tonic, giving tone and healthy action to the emunctories. 3rd. As a refrigerant; while the purgative articles combined with it expel the morbid secretions, and disgorge the glandular apparatus. I can only . . . add, that three years experience does not enable me to say, that I have ever seen the first symptom of colic, or the least unpleasant consequence follow its application; but on the other hand, I have had every reason to place the most implicit confidence in its powers.[48]

Today lead as a therapeutic agent, employed externally or internally, has become a thing of the past. This is attributable to advanced knowledge of the toxicity of the metal in its various forms when introduced, either directly or through absorption, into the human system.

Manganese

> An article published in the MEDICAL GAZETTE, Nov. 8th, of the past year, entitled "Researches on Gout," by a gentleman eminently celebrated for his knowledge of chemistry, and his application of that knowledge to pathology and therapeutics (Mr. Alexander Ure), called my attention to the use of the sulphate of manganese. The salt has given some marked results affecting the biliary secretions in a remarkable degree. Under the impression that it may make a useful addition to our Pharmacopœia, I have been induced to offer you the results of my experience in its use, claiming no further merit in its introduction than an honest desire to test the truth of Mr. Ure's suggestion by such experience as has been afforded me. When taken upon an empty stomach, in doses of one or two drachms, it has invariably produced vomiting in less than three hours, and generally within an hour; and the matter vomited has consisted of a very large quantity of yellow bile. After a meal, the same effect has taken place, but not invariably.
>
> It very rarely acts as a purgative alone, and after it has been exhibited for several days, I have often been obliged to have recourse to other purgative medicines, in consequence of the want of action of the bowel. After the first dose it seldom acts as an emetic. The appetite has invariably increased during the exhibition, and when the first emetic effect has subsided the patient is free from all uneasy sensations, and expresses himself as feeling lighter and easier than before.[49]

This report was made in 1844. A decade later an American professor of materia medica, E. H. Davis, complained of the neglect of salts of manganese by the profession. The therapeutic value of manganese had been recognized in Mâcon, France, where workers in the manganese mines "were uniformly cured of scabies and other cutaneous affections during their stay at the works." This led to general use of the oxide in such cases.[50]

Further investigation by the French led to the discovery of the presence of manganese in the blood and to the conclusion that the metal was an acceptable, sometimes superior, substitute for iron in the treatment of anemia. In fact, manganese offered the advantage over iron "that its preparations may be combined with all the vegetable tonics and astringents without risk of chemical incompatibility." [51]

By the 1880s a fourth medicinal use had been found for manganese salts. This was in the treatment of amenorrhea (absence or suppression of menstrual flow). On 5 January 1885, Franklin H. Martin (1857–1939), who would subsequently (1905) found and edit *Surgery, Gynecology, and Obstetrics,* read a paper before the Chicago Medical Society in which he summed up the findings to date, on manganese as a stimulant to menstrual function, of such authorities as Sydney Ringer (1835–1910) of Norwich, England, Sir William Broadbent (1835–1907) of London, T. Gaillard Thomas (1831–1903) of New York, and Roberts Bartholow of Cincinnati and Philadelphia.

Martin had published the results of his earlier investigations in the *New York Record* (29 September 1883) and now wished "to give additional testimony confirming the efficacy of the remedy." He had found that, besides relieving certain forms of amenorrhea, manganese was valuable in the treatment of menorrhagia (excessive bleeding at the menstrual period) and metrorrhagia (bleeding at times other than the menstrual period). "From my observation I have been led to consider manganese in any form a direct stimulant to the uterus and its appendages."

Permanganate of potash was the salt usually administered. Martin pointed to its "disagreeable, distressing effect on the stomach when taken undiluted, which may be obviated by administering on a full stomach—immediately after eating—or dissolved in considerable water." He had found dry gelatine capsules to be the most convenient form in which to administer the preparation.[52]

Nickel

> As with all else that is new, a certain scepticism has ever prevailed concerning the utility of new drugs. Such conservatism is doubtless beneficial, for how can any good come of changing an old drug of known utility for a new one of doubtful worth? Nevertheless, many remedies of great utility receive frequent condemnation at the hands of the profession simply because the proper conditions indicating their use are not known or appreciated.[53]

In the 1870s Da Costa introduced the bromide of nickel as a useful remedy in the treatment of epilepsy and kindred diseases. It was not until 1884 that extensive trial of the bromide was made in the outpatient department of the Jefferson Medical College Hospital, Philadelphia. It was at first found that when this salt of nickel was administered, while some cases benefited greatly, in others less relief was obtained than from the combined use of the bromides of sodium and potassium. In time it became clear that good results from nickel occurred in cases where attacks took place regularly and at comparatively long intervals. When attacks were frequent (several in twenty-four hours) better results came from the use of the other bromides. "In other words, where the object in view is to keep up a mild impression for a long while, that then the results obtained from the bromide of nickel are most gratifying." But would not small doses of the other bromides be equally satisfactory? The answer was no, if only for the reasons that the bromide of nickel disordered the digestive tract less than any of the other bromides and was much less of a depressant to the nervous system. For the latter reason, bromide of nickel was especially indicated when a hysterical or hypochondriacal element was prominent in a case.

Experimentation at Jefferson clearly showed that, in addition to being effective in the treatment of two general classes of epilepsy, the bromide of nickel would relieve congestive headache and was an excellent remedy for wakefulness resulting from long-continued excitement of the nervous system. "We claim for this drug no specific curative action; we only wish to state our belief that it is a most valuable addition to the therapeutics of epilepsy, . . . and also a potent remedy in many kindred disorders." [54]

10

The Earth Metals

There are two groups of earth metals. The alkaline earth metals, which do not occur free in nature, are good conductors of electricity, and are about as hard as lead. Among them barium and calcium were of medical importance prior to the twentieth century. (Radium, properly speaking, is an alkaline earth metal but has generally been considered separately because of its radioactivity.) The other group, the rare earth metals, has, with the exception of cerium (isolated in 1809), only recently become available for practical use.

Barium

In 1784 Adair Crawford (1748–1795), physician to St. Thomas's Hospital, began investigating the medicinal properties of muriated barytes (chlorinated barium) and arrived at the conclusion that "it might possibly possess considerable powers as a deobstruent."

> This salt, when perfectly neutral, has a bitter taste; but the portion of it which was used in my first experiments happened to contain a small excess of acid, by means of which the bitterness was in a great measure destroyed, and its taste became somewhat similar to that of common salt. From this similarity it seemed not impossible, that its virtues might resemble those of the latter substance; and there could be little doubt that its activity would be greater, because the heavy earth, in several of its properties, has a considerable resemblance to metallic calx.
>
> I moreover found, that a small quantity of the muriated barytes, when it was dissolved in water, and taken into the stomach, excited an agreeable sensation of warmth; and I had reason to believe that it would act gently as a laxative and a diaphoretic.
>
> These facts led me to suppose that it would make a valuable addition to the materia medica. I was, however, at that time obliged, by my other avocations, to omit the further prosecution of this subject. But last winter [1788–89] having acquired more leisure, I determined to resume it, and to give the salt a fair trial in scrophulous and cancerous complaints.

Crawford reviewed seventeen cases he had under treatment and observation and reached these conclusions:

> It appears, in general, that very little relief was afforded by [muriated barytes] in the last stages of cancer and consumption. But in all other cases in which it was tried, its exhibition was evidently productive of salutary effects.
>
> Indeed, in some instances, it removed diseases, which, I believe, could not have been subdued by any other remedy; particularly in scrophulous complaints, in which it seems to have acted with a degree of force and certainty hitherto unexampled in the records of medicine.

When this remedy was given in a moderate dose, it appears, in a few instances, to have increased the secretion by the skin; in a great variety of cases it occasioned an unusual flow of urine, and it almost universally improved the appetite and general health.

It seems, indeed, to combine within itself the qualities of an evacuant, a deobstruent, and a tonic. I have sometimes observed, that it occasioned vertigo. This effect I ascribed in some measure to the nausea which it excited. Like every other active medicine, it would, no doubt, if administered injudiciously, be capable of producing deleterious effects.

In a considerable dose, frequently repeated, it would lessen the appetite, by the constant sickness of stomach which it would occasion; and in a still greater dose, it might be productive of much danger, by disordering the nervous system, and by operating violently as an emetic and purgative.

It is proper to remark, that the salt that was exhibited in the foregoing cases previously to the beginning of May, did not consist of the muriated barytes in a state of perfect purity. . . . I have, however, sufficient reason to be convinced that the [muriated barytes] is a very efficacious medicine.[1]

It will be recalled that Humphry Davy in 1808 was attempting to extract barium from its alkaline earth.

Crawford warned that from "trials which have been made with dogs it appears, that a very large dose of muriated barytes would prove fatal. I therefore think it necessary to caution those who are unskilled in medicine, not to tamper with this remedy." [2] Time has revealed that all soluble salts of barium are exceedingly poisonous. Consequently the only form of barium in use today is the sulfate, which is insoluble in water, organic solvents, and aqueous solutions of acids and alkalies. It is employed in radiography of the organs of the gastrointestinal tract, particularly in the diagnosis of peptic ulcer, carcinoma, diverticula, and adhesions.

Calcium

In 1689 Walter Harris, in a book on infants' diseases that remained authoritative for a century, emphasized the use of calcium in the treatment of infantile convulsions (which were still confused with epilepsy). In the middle of the eighteenth century, Jean Astruc (1684–1766), professor of medicine at the Collège Royal de France in Paris, recommended the treating of rickets with solutions of calcium and sodium phosphates, but few subscribed to this approach.

Around 1750 Lorenz Heister (1683–1758), professor of surgery at the German University of Helmstedt, advocated the cleansing of infected wounds prior to treatment, recognizing calcium hydrochloride as a beneficial detergent. In 1787 the Dublin obstetrician Joseph Clarke (1758–1834) used calcium hydroxide as a disinfectant to speed recoveries in cases of puerperal fever. Antoine-Germaine Labarque (1777–1850) employed it as a surgical disinfectant.

Muriate of lime (calcium chloride) appears to have been earliest employed as a therapeutic agent by the distinguished French physician Antoine-François de Fourcroy (1755–1809) who, with some Dutch and German physicians of his day, had much confidence in its powers over scrofula. By the beginning of the nineteenth century the remedy had become highly prized in Britain.

> The want of success which had attended the use of calomel, sponge, steel, Peruvian bark, tepid salt-water bathing, muriate of barytes, and all other remedies which were commonly employed had led Dr. [Thomas] Beddoes [(1760–1808) of Shifnal, Shropshire, the sponsor of Humphry Davy,] to make trial of a remedy which, to use his own words, "was strongly recommended in scrofula by some foreign writer, the muriatic acid saturated with lime, or muriate of lime, as it is now styled." Dr. Beddoes gave

the remedy to nearly one hundred patients in various conditions of life . . . [and found] that there are few of the common forms of scrofula in which he had not had successful experience of the medicine.[3]

Beddoes's observations were published in the *Annals of Medicine* in 1801. A few years later the first issue of the *Edinburgh Medical and Surgical Journal* carried an article by James Wood, physician to the Newcastle (England) Infirmary and Dispensary, in which he had this to say:

In my inaugural dissertation on scrofula, published in 1791, I endeavoured to show, from the experience I had had in that disease, that it was entirely a disease of debility, to be prevented and cured by those means which would prevent or remove that debility. At that time, though the tonics, with which we were acquainted, such as cinchona and steel, had the good effects to be expected from them, yet they were neither decisive at the moment, nor permanent; they only seemed to suspend the disease, until some change of the constitution of the patient removed the scrofulous action: and though, in the same dissertation, I mentioned my experience of the good effects of the aqua calcis [lime-water], as a remedy in this disease, yet I could not ascribe to it, given separately, greater virtue than to cinchona, or other tonics, although I mentioned having perceived more evident advantages from them when conjoined, than from either alone. Cinchona bark infused in lime water . . . were the forms I used.

Very soon after the period I have mentioned, the muriate of barytes was recommended as a remedy in scrofula. . . . While experiencing good effects from the muriate of barytes, the muriate of lime was recommended to my notice . . . as possessing great powers in discussing tumours and obstructions of different kinds. The use of the remedy was confined, for some time, to particular species of obstructions; and afterwards, when given in scrofula, it was commonly united with decoction of cinchona bark. It was not, therefore, until the beginning of the present

> year (1804), that I determined to make trials and experiments with the muriate of lime by itself, and more particularly in the cure of scrofula. These trials were made on a very extensive scale, and they were immediately decisive in its favour. The muriate of lime produced all the good effects which I had experienced from the muriate of barytes, with two great advantages in addition; its action was more immediate, and no bad consequences attended an over-dose.[4]

Wood used calcium chloride in the treatment of incipient tuberculosis of the lungs, in all forms of scrofula, and in rickets. In 1808 James Sanders (1777–1843) of Edinburgh made this positive statement: "I think that I have ascertained that the muriate of lime has a more powerful effect in removing indolent scrofulous tumours than any other substance used as a remedy, but that when they become open sores it is almost useless." [5]

Calcium chloride had other supporters in James Hamilton (d. 1839), professor of midwifery at the University of Edinburgh for most of the first half of the nineteenth century, Anthony Todd Thomson, and Joseph de Vering, who published a book on the cure of scrofula in 1832.

But it also had its detractors. John Thomson (1745–1846) had this to say in a treatise published in 1813:

> Three of the neutral salts have acquired great celebrity for the cure of scrofula; and it is remarkable enough that these three should all have been muriates. The first of these was muriate of soda, given as it exists in sea water. Nothing can be more satisfactory than the evidence which is on record of its efficacy. In reading this, one only wonders how so efficacious a remedy should ever have fallen into neglect. The second, the muriate of barytes, was introduced to the notice of the public under the most favourable auspices, and its antiscrophulous powers extolled by all degrees of men in the medical profession; yet it has had a much shorter-lived reputation than sea water, or its successor, the muriate of lime. How long this third muriate will be permitted

> to enjoy its present fame, I shall not venture to say. Not much longer however, I should imagine, from what I have seen of its use, than till a new remedy shall be found out by those who are still sanguine in their hopes of discovering a specific for scrophula. To such of you as are but imperfectly acquainted with the past history of the materia medica, and the uncertain nature of medical evidence, in so far as it relates to the operation of remedies for the cure of chronic diseases, the accounts which are already before the public, of the virtues of muriate of lime in curing scrophula, must appear most satisfactory and complete. It will be well if a little reading or experience does not soon lead you to suspect, that the reporters of its efficacy have not, any more than the reporters of the efficacy of the muriates of soda and barytes, learned to distinguish, in every instance, between a *cure* and a *recovery*. Till that distinction however is made, and is adhered to more strictly than appears to have been hitherto done in reporting the effects of the remedies employed for the cure of scrophula, a little scepticism, even with regard to the anti-scrophulous virtues of muriate of lime, may, I conceive, be safely enough indulged.[6]

Benjamin Phillips, author of *Scrofula: its Nature, its Causes, its Prevalence, and the Principles of Treatment,* expressed himself as "not satisfied that it has any very evident action upon scrofulous glands. I cannot say that I have ever seen a case in which, in the absence of other influences, the discutient power of this medicine has been clearly manifested."[7] Samuel Cooper (1780–1848), in the sixth edition of his *Dictionary of Practical Surgery* (first published in 1809), wrote: "I have seen the muriate of lime given in several cases of scrofula, but without any beneficial effect on the disease."[8]

Having reviewed the pros and cons in a paper read to the Medico-Chirurgical Society of Edinburgh on 15 May 1872, J. Warburton Begbie (1826–1875), second son of James, was of the opinion that the former outweighed the latter, but he was forced to admit that "the adverse opinion respecting the use of muriate of lime led in some measures at least to it being less employed as a remedy, admits of little doubt. . ." [9] However, he preferred to attribute the fact that calcium chloride, after being in high esteem as a remedy during a considerable period of time, had of late years passed almost entirely into disuse, to the introduction of iodine and cod liver oil. On the other hand, James Wood was seemingly in error when he declared that "no bad consequences attended an over-dose."

> Chlorine of calcium is, in large doses, an irritant poison. . . . When exhibited in large doses to man, the muriate of lime excites nausea, vomiting, and sometimes purging, causes tenderness of the precordium, quickens the pulse, and occasions faintness, weakness, anxiety, trembling, and giddiness. In excessive or poisonous doses, disorder of the nervous system is manifested by failure and trembling of the limbs, giddiness, small contracted pulse, cold sweats, convulsions, paralysis, insensibility, and death.[10]

In 1855 John Cleland, demonstrator of anatomy at the University of Edinburgh, saw (at the École de Médecine in Paris) saccharated lime (calcium gluconate) used in determining the amount of nitrogen in organic substances. "It then occurred to me that this solution would be a useful agent in medicine; for it was evident that, while the lime-water in use was far too weak a preparation to develope to advantage the therapeutic properties of lime, its utility was such as to render it highly probable that a sufficiently strong solution would be at once valuable as a tonic and antacid." Cleland made his first trials of the medicinal effects of the gluconate in the winter of 1856 and 1857.

> It is of course a powerful antacid, and probably the best we have, since it is stronger and pleasanter than magnesia, and does not weaken digestion like the alkalies. Far from doing so, its most

> important use is as a tonic of the alimentary system in cases of obstinate dyspepsia. As such, its action is much more powerful than that of the vegetable stomachic tonics. It is suitable for cases with too little as well as for those with too great secretion of gastric juice, no doubt because the former state of matters is obviously a result of atony, which the lime removes. It seems particularly serviceable in gouty constitutions.[11]

At the time Cleland claimed to be the first to introduce the gluconate as a medicine, but in a further paper published in 1875 he conceded: "I am led to understand that I was not the first to do so, and that it had been previously recommended by Dr. Capitaine."[12]

It came to be recognized that calcium gluconate was as effective as calcium chloride in raising the calcium content of the blood and was at the same time more pleasant to take (being odorless and tasteless instead of having a sharp, saline taste), and nonirritating.

Arts Revealed and Universal Guide, a family handbook, published anonymously in Indianapolis in 1859, offered a single cure (which turned out to be calcium phosphate and cod-liver oil) for consumption, scrofula, general infantile atrophy, rickets, diarrhea, and tuberculosis.

A decade later the medicinal value of the phosphate in combatting fevers and speeding the subsequent convalescence was investigated.

> In a great number of acute diseases, especially in typhoid fever, typhus, and certain forms of pneumonia, we often notice a condition of considerable adynamia [weakness] which takes its origin either in the peculiar character of the malady or in the constitution of the patient, and is consequently attended by a marked rise of temperature. . . . Every substance whose action tends to induce a sedation of [the ganglionic] nervous system may therefore,

> by suspending or diminishing combustion, lower the temperature and oppose the progress of spoliation and of weakening of the organism. . . . But even where the acute condition of the disease no longer exists, there subsists during the period of convalescence a general atony proportional to the gravity and duration of the febrile state which had brought on a period of arrest in the phenomena of nutrition of the tissues: several months may then pass on before health is re-established, if, indeed, it is ever entirely recovered. . . . With the view of warding off such results, I put the question whether it would not be more rational to resort in such cases to agents capable of arousing in the organism the entire energy of the nutritive function, and this substance, I believe, is to be found in the phosphate of lime, combined with lactic acid, and already known in France by the name "lacto-phosphate of lime." . . . The lacto-phosphate of lime is at once an aliment and an article of food, and a medicament of the highest value. Its administration cannot, like that of alcohol, produce mischievous consquences, and it never depresses the nervous system charged with presiding over the transformations which take place in the nature or form of the elements of our tissues.[13]

The Irish pathologist Almroth Edward Wright (1861–1947) has been credited with having in 1891 defined the role of the calcium salts in the coagulation of the blood, but the French physiologist Nicholas Maurice Arthus (1862–1945), working independently, made the same discovery and published his findings a year earlier.

Cerium

In 1854 James Young Simpson (1811–1870), professor of obstetrics at the University of Edinburgh, drew the attention of the profession to the value of the oxalate of cerium in combatting nausea and vomiting associated with pregnancy. He considered it a sedative tonic that compared favorably with the salts of bismuth, for which he believed it should be substituted. Six years later Charles Lee, house physician to Blockley Hospital, Philadelphia, reported that he had successfully extended the use of cerium to stemming vomiting accompanying pulmonary consumption (phthisis) and severe heartburn (pyrosis) and to combatting hysterical vomiting and lack of muscular tone in the digestive organs (atonic dyspepsia). He was singularly impressed by the rapidity of the oxalate of cerium's therapeutic action and was of the opinion that it should assume a permanent place among the mineral tonics.[14]

11

The Alkali Metals

The alkali metals, of which sodium and potassium are prime examples, are never found uncombined in nature; having no utility as structural metals, they are largely employed as chemical reactants. The exception is lithium, which tends to resemble the alkaline earth metals.

Potassium and Sodium

Among the ancient Greeks and Romans *niter* meant either carbonate of sodium or carbonate of potassium (usually the former). To Arab chemists—and to Roger Bacon—the word *niter* always meant nitrate of potash. Around 1672, Peter Seignette (d. 1716), a French apothecary, was making tartrate of potash (soluble tartar) from cream of tartar and carbonate of potash when he inadvertantly added carbonate of soda. His "new discovery," Rochelle salt, named for his place of residence, is employed to this day as a saline cathartic—potassium sodium tartrate. Humphry Davy devoted his 1807 electrochemical researches to potash and soda. In short, the metals potassium and sodium, in their various forms, have been linked together—even mistaken one for the other—down through the ages, and the confusion that has sometimes resulted dictates their combined consideration.

At the beginning of the nineteenth century it was believed that no one could recover from a large dose of potassium nitrate. Andreas Comparetti (1746–1801), known for his work on the morphology of the ear, had reported the case of a man who died as a result of accidentally taking an ounce and a half of niter in solution. Orfila, in whose opinion niter acted on the stomach as an acrid and corrosive poison, had related many cases where even an ounce or less of niter had destroyed an adult, while offering no instance of recovery.

On 17 March 1815 the twenty-five-year-old wife of the quartermaster of the South Devon militia swallowed two ounces of dissolved niter, mistaking it for an ounce of Epsom salt. Vomiting followed almost immediately. She brought up the contents of the stomach and then blood. At the sight of the blood a neighbor sent for John Butter (1790–1876), surgeon to the South Devon militia. Progress was slow, but Butter pulled her through, and on 31 October, although "her labour was more tedious than usual, owing to debility," he delivered her of a fine male child, and "on the 29th October 1817, I attended her during the birth of another son, all of whom are living and well." This case not only showed the quantity of niter the stomach could tolerate under proper treatment but proved that the most violent emetics do not necessarily produce abortion. In conclusion, Butter ranked niter among "those poisonous substances, which, the more dangerous they are, become the most useful in skilful hands." [1]

> Nitrate of Soda dissolves the protein element of the blood much less than Nitrate of Potassa, coagulated fibrin being but very little or rather not at all influenced by it, while at the same time it contracts the blood corpuscles much more firmly even to shriveling, and renders the serum redder and richer in hæmatin.[2]

J. B. Brown, in an article published in 1854, credited Dr. Johann Gottfried Rademacher (1771–1850) of Vienna with being the first to use sodium nitrate extensively as a therapeutic agent.

> As he attributes to it the most extraordinary properties in different diseases, claiming it as a universal remedy and recommending in the most heterogeneous affections without stating any particular indications for its use, I was induced two years since to make some experiments myself for the purpose of ascertaining whether the medicine possessed the value which its friends claimed for it. Rademacher thinks it is more useful in gastric fevers than Nit. Potassæ, which as an active solvent of fibrine, causes a more rapid putrescence, and which in cases where the a priori zymotie [sic] tendency is evidently contraindicated, whilst the Nit. Soda although it restrains the pseudoplastic process, does not produce any excessive evacuations.[3]

Brown found that sodium nitrate given as a gentle laxative was very acceptable but that large quantities produced painful spasmodic contraction of the anal and bladder sphincters. Nevertheless, he found it superior to all other agents in the treatment of acute or chronic dysentery.

> I have been now in more than a hundred cases of dysentery so successful, as to restore my patients within the period of from 4 to 14 days at most, under the principal influence of Nitrate of Soda, never using a single grain of calomel, which I avoid for fear of salivation and of its consequence, these being often as bad a

> complaint as the former, or even still worse, I would therefore recommend this salt as a very powerful and innocent substitute for calomel in all the different forms of dysenteric and congestive diarrhœa and dysentery.[4]

In 1873 H. Macnaughton Jones (1844–1918) of Cork Fever Hospital, who produced an *Atlas of diseases of the membrana tympani* (London, 1878), reported that he had, with almost universal success, employed potassium nitrate in diseases where the temperature maintained a high range, including pneumonias.[5]

The iodohydrargyrate of potassium (prepared by combining pure iodine of potassium and the dentiodide of mercury in a small amount of water) was discovered in 1826 by P. A. de Bonsdorf of the University of Finland. Within seven years it had come to be regarded as something of a panacea by William Channing of New York. A review of cases in which he had employed the agent led him to conclude that it had beneficial effects upon the organs of circulation, the lungs, the alimentary canal, the urinary organs, the skin and cellular tissue, and the absorbent and exhalant systems. Diseases that had uniformly reacted better to its use than to any other treatment were chronic bronchitis, "hooping-cough," pulmonary consumption, thrush, tonsillitis, pharyngitis, chronic gastroenteritis, colitis, uric acid crystals, diabetes, excessive bleeding at the menstrual period, absence or suppression of menstruation, infection of the vagina or cervix, herpes, psora, psoriasis, severe generalized edema, suppurative inflammation of the eye, and scrofula. Channing believed that the agent was no less effective for another group, though his experience was "too limited to authorize assurance." This group involved peripneumonia, pleuritis, hemorrhoids, intestinal worms, hepatitis, peritonitis, inflammation of the kidney, mucous discharge in chronic gonorrhea, chronic eczema, scalp disease, leprosy, and carcinoma. Channing added that gonorrhea and syphilis had been successfully treated but "it is believed that other means are to be preferred, excepting when these affections occur in scrofulous [tuberculous] habits."

> If this wide range of disease, alarm the incredulity of the cautious practitioner, perhaps his faith may be revived when he shall call to mind, not the fabled virtues of a panacea too often practically assigned to mercury, but the *well-authenticated facts,* showing the unrivalled efficacy in a long catalogue of diseases, of the several elements here associated in chemical combination. So far then from shaking his confidence, should not this consideration rather urge him to subject to the test of clinical experiment the remedial powers of the article, and determine for himself its real value as a therapeutical agent.
>
> That an instrument of such potency will be exposed to the abuses incidental to ignorance and empiricism, is sufficiently indicated by the history of tartarized antimony, of quinine, and of every valuable accession to the materia medica. But the mischief wrought by such weapons in the hands of the charlatan, will never deter the scientific physician, whose skill knows how to wield them efficiently in behalf of suffering humanity.[6]

In 1836 William Wallace (1781–1837) of the Dublin school introduced the use of potassium iodide in syphilis. Twenty years later an English physician, William O'Connor (d. 1880), was the first to use potassium iodide in the treatment of epilepsy.

In 1875 Joseph R. Beck of Fort Wayne, Indiana, reported the successful use of the iodide not only in syphilis but in all forms of rheumatism, including gout. By 1884 H. W. Berg of the New York Orthopedic Dispensary could write: "Amidst the prevailing scepticism in therapeutics, few remedies have better maintained the right to be considered curative agents than iodide of potassium in the treatment of tertiary syphilis." Berg went on to regret, however, what he admitted to be true of all drugs, that, "in addition to its curative effects, [there are] others, which are not only non-curative, but positively disadvantageous and injurious." He then proceeded to weigh the potential advantages of sodium iodide over potassium iodide in the treatment not only of syphilis but also chronic rheumatism and concluded that sodium iodide could be used therapeutically with as beneficial effects for the chief, if not all, purposes for which potassium iodide was employed; that sodium iodide was more assimilable than the iodide of potassium, both in the digestive organs and in the general system and that, as a consequence, many of the local and general undesirable effects of potassium iodide did not follow.[7]

Potassium chlorate was first used in medicine toward the end of the eighteenth century, but it was not well known to the profession until the mid-1800s. By 1845 obstetrician James Y. Simpson was employing the salt with seeming success in cases where patients had lost the children of previous pregnancies from disease of the placenta. He had found that a clot of newly drawn venous blood covered with a thin layer of water retained its dark color but that when an alkali salt was added the clot very quickly attained a red arterial color. In feeble placenta cases he therefore recommended the employment of such a salt to render the blood more arterial, and he chose the chlorate of potassium because of its high oxygen content.

Inspired by having as a student seen the drug employed in two cases, Alexander Russell Simpson, nephew of the celebrated obstetrician, subsequently professor of midwifery and the diseases of women and children at the University of Edinburgh, decided in 1857 to investigate potassium chlorate. Ernst Felix Immanuel Hoppe-Seyler (1825–1895) of Virchow's Pathological Institute at Berlin (where Simpson was then working) introduced him to the work of Emile Isambert (1827–1876), whose definitive memoir on the subject had been published in Paris a year earlier. Isambert showed that, within a short time after the chlorate had been administered by mouth, it could be discovered in almost all the secretions of the body. "I was able to confirm Isambert's observation that a careful quantitative analysis of the urine passed within thirty hours after the administration of a dose of chlorate, yields such a proportion as practically represents the whole of the drug." [8] Nonetheless, Isambert found, potassium chlorate augmented appetite, provoked salivation, increased the flow of bile, and was (in large doses) a diuretic and a sedative to the action of the heart.

In an article published in 1888 Simpson reviewed the literature of three decades and concluded that two kinds of cases seemed to benefit from the use of the drug. "These are, *first,* women who have lost a series of children from syphilis; and *second,* those whose infants have died with placental apoplexies, usually associated with albuminuria." [9]

In 1856 an English surgeon named Hutchinson, at the Metropolitan Free Hospital, reported the successful employment of potassium chlorate in a number of cases of ulcerations or cancer sores on the mucosa of the mouth, in both adults and children, and in cases of excessive secretion of saliva following mercury treatment for syphilis. By 1860 E. J. Fountain of Davenport, Iowa, had extended use of the chlorate to phthisis, scrofula, and other diseases.

> Believing the chlorate of potash to be a *tonic, alterative and blood depurant* superior, in many respects, to anything in our materia medica, I do not hesitate again to urge its claims upon the attention of the profession. From time to time when, in 1851, I discovered its utility as a remedy for mercurial ptyalism, I have been testing its properties in many directions, and with constantly increasing confidence in its virtues, and (in the language

> of Dr. Hutchinson) "its scarcely less than wonderful power." I believe there is no tonic and alterative comparable to it in all manners of scrofulous diseases, and every form of tuberculosis, in the sequelæ of the exanthemata [skin eruption with inflammation], in adynamic fevers, and asthenic types of disease with depressed vital power, in every direction involving an imperfect aeration of the blood, and many derangements of the system resulting in abscesses, eruptions, ulcers, &c. In my opinion, it has this wide range in its application by reason of its *oxydizing* properties, in support of which I may have something to say hereafter.[10]

As with the iodide, there were those who preferred the chlorate of sodium to that of potassium. In May 1884 Trail Green of Easton, Pennsylvania, told the Section of Medicine and Materia Medica of the American Medical Association: "I may say that there are no cases in which the potassium salt has been found useful that will not be more rapidly cured by the sodium salt." [11]

Bromine was discovered in 1826 by Antone Jerôme Balard (1802–1876) of Montpellier, who developed the bromides of potassium and sodium. According to Karl Binz (1832–1912) of the University of Bonn, potassium bromide was first employed therapeutically ten years later by a Dr. Pourché, also of Montpellier.[12] The next thirty years saw considerable research involving animals and men, and by 1868 potassium bromide was generally accepted as a nerve sedative, although it was considered to be effective in small doses only. According to Wilhelm Sander (1838–1922), a psychiatrist of Dalldorf, "If we give, at first, 78 grains daily (in 3½ ounces of water only), and gradually increase the quantity to half as much again, there occurs in many, but far from universally, an improvement in so far as that the number of the (epileptic) attacks diminishes, and the individual fits abate of their intensity and duration. In some cases, in which the regular and typical occurrence of the attacks had been watched for a long time before the bromide was given, the fits remained absent, under the remedy, for several weeks or even months." He pointed out, however, that the chloride of potassium gave the same results in epilepsy and that he had not seen any case in which the chloride failed and the bromide subsequently succeeded, concluding that the chloride possessed the advantages that there were no serious side effects (such as the acne-like eruption produced by the bromide), that smaller doses were sufficient since the chloride contained more potassium than the bromide, and that it was considerably pleasanter to take.[13]

In the meantime, James Begbie had set out in 1866 to answer a question, recently "raised and partially discussed, in the pages of periodical literature, in regard to the efficacy of the bromides of ammonium and potassium as remedial agents in disease."

> The answer, I apprehend, has been disparaging and unjust to their value, and calculated to deter the profession from entering on a full inquiry into their real therapeutic properties. After an experience of some years of the use of these remedies, particularly that of the bromide of potassium, I have become satisfied of their great value in the treatment of many diseases, but more especially of disorders of the nervous system,—affections of centric origin or of remote parts through reflex action.[14]

Begbie's descriptions are sometimes picturesque:

> The bromide of potassium is a valuable calmative and hypnotic. When opium and other narcotics have failed to produce sleep, or when they have succeeded only at the expense of sickness, vomiting, headache, and other consequences, the bromide, free from all unpleasant or injurious effects, will often tranquillize and secure repose. . . .
>
> In those distressing nervous affections, the offspring of over-

taxed brain, which we are ever and anon called upon to combat in the case of the earnest student, the plodding man of business, or the speculating merchant, cases where, by rising early, and sitting up late, neglecting regular hours of diet, and abandoning exercise in the open air, the whole machinery of life and health have been deranged, and the unhappy victims contemplate nothing short of the wreck of mind and body: in these circumstances, next to rigid hygienic rules imposed by the physician, and carefully carried out by the patient, will be found the amelioration and ultimate removal of the evil, in the use of the remedies which have a calmative effect upon the nervous system. Of these the bromides, in my experience, are the safest and the best. . . .

In those sad and humiliating instances of nervous disorder which follow the vicious practice in which unhappily the young, and sometimes the more mature of both sexes indulge, the painful retribution which sooner or later overtakes them is often the call for medical assistance. In opium and belladonna, in zinc and iron, there is often found a source of relief so far as medicine is capable of affording it. In the bromides we secure an additional and trustworthy agent, of whose efficacy I have repeatedly satisfied myself.

Closely allied to, and springing out of, the nervous disturbance produced by the practice referred to, are various shades of epileptiform disorder, and even epilepsy itself in no mitigated character. Of the powerful influence for good which the bromides have exercized over such affections there can be no doubt. . . .

Opium, antimony, aconite, digitalis, and others have been employed in large and frequently repeated doses, as powerful depressants in the paroxysms of acute mania and delirium tremens. The practice is not without its risks. In securing the calmative and sedative effects of the bromides will be found a safer and, I venture to hope, a not less successful mode of practice. . . .

Such are some of the effects of a remedy which is destined, I believe, to hold its place among our valued therapeutic agents; in its mode of operation to throw light on some obscure affections;

> to afford comfort and relief to many suffering from painful disorders; and to falsify the prediction, too readily announced, that "bromine and its compounds are already sharing the fate of many of their predecessors, and falling gradually into the sere and yellow leaf of fashion, as a prelude to their being entirely consigned to oblivion." [15]

Sander, in advocating large doses of potassium bromide, proposed seventy-eight grains daily, gradually increased to half as much again, but *The British Medical Journal* for 16 October 1869, carried an account by a patient (not a member of the profession) of the ill-effects of dosages in this range:

> It was in June 1867 that I began taking the bromide; the daily dose being then, I think, about twenty grains. It very soon caused the cessation of the "lapses" (*petit mal*); and, in order to make sure and stop the greater evil also, I went on increasing the dose (hardly with Dr. M.'s permission, and yet not against his orders) till at length I should think I must have been taking seventy grains a day, perhaps sometimes eighty. The first symptom of overdoing the thing that I noticed was the profound and yet disturbed sleep into which it seemed to throw me. I always awoke with a mental struggle and effort, not knowing at first where I was, or what had become of me; in fact, as I told Dr. M., I seemed to have gone too far down into the gulf of sleep. Side by side with this but, of course, less noticeable to me, was the enfeebling of mental power. A little page in my accounts, which I should usually prepare and balance in half an hour, took me two or three evenings' weary work. But the worst thing was the tendency to talk "Mrs. Malaprop" English, substituting one word ending in "tion" for another in a most provoking and yet ludicrous way. I had once to write some letters reminding people that their subscriptions were due, and I had the misfortune of having my letters (I think one or two of them) brought back to me by a clerk, who pointed out to me that I had written "contraction", or some such word, instead of "subscription". I cannot

just now remember any more instances; but this difficulty in getting and keeping the right word (though the right idea was present in my mind) is very vividly, and not without humiliation, present to my recollection. Soon, of course, my wife and partner saw the change in me, and attributed it to the right cause. I went from home, and for a time dropped the medicine. In a week, my host said, "Why, you look ten years younger than when you first came." The stoop in my figure, the slow uncertain speech, and other bad symptoms, especially the heaviness in the eyes, were gone, and I felt quite myself again. I am still taking the medicine, but now never exceed forty grains a day, often taking only twenty; and, if I find the slightest touch of the "Mrs. Malaprop" difficulty, I reduce the dose at once.[16]

Potassium bromide was faring better in America; Lunsford P. Yandell, Jr., professor of materia medica and clinical medicine at the University of Louisville, could report, also in 1869:

No medicine, at the present time, is probably attracting so much attention as the bromide of potassium. Besides being a very fashionable drug with the profession, it has become a popular favorite, and now occupies the same rank in domestic medicine that the tincture of arnica has so long held in domestic surgery. While a very wide difference of opinion exists among writers and practitioners in regard to its curative powers and its behavior in the human system, the mass of testimony is undoubtedly in its favor. During the past two years my experience with it has been somewhat extensive both in hospital and private practice; and the result of my observations is, that while it is by no means the panacea which some have deemed it, it is a most valuable medicine, capable of doing great good in many morbid conditions, and nearly destitute of all power of evil when properly administered, even in very large doses.[17]

Yandell had successfully employed the bromide in cases of locomotor ataxia, aphasia, epilepsy, sick headaches, pure neuralgia, cerebrospinal meningitis, and in helping alcoholics to taper off.

Binz's article was a careful review of the literature covering bromide of potassium for the years 1861 through 1873, but Francis Edward Anstie, editor of the *Practitioner* (in which Binz's article appeared) commented in the same issue that the "severely sceptical paper of Professor Binz which precedes this article will be read with surprise by most English medical men . . ." [18]

Binz's "scepticism"—it almost ranks as prejudice—is well illustrated by the following passage in respect of Hermann Senator (1834–1911) of Gnesen, Polish Prussia, whose reputation was built on his investigation of the pathology of fever and its treatment. Binz quoted Senator's "richly stored book 'On the febrile process,' " (Berlin, 1873) as follows: "There is only one potassium-salt from which, according to my experience, we may with security expect a direct and most desirable action in febrile diseases, viz., the bromide of potassium, which, appropriately administered, subdues the insomnia and restlessness better than any of the medicines ordinarily given." He then added: "The author is accustomed to the exact method, and we have no grounds for doubting the correctness of his observation. However, even here the question remains open as to the specifically active element of the drug." [19] But on the subject of specificness he chose to ignore an earlier statement of John Russell Reynolds (1828–1896), well-known author of standard works on epilepsy:

> It is to be demonstrated, in my opinion, that there is something "specific" in the action of KBr. Potassium—given as iodine—is without such effect; and bromine—given as bromide of ammonium—has no obvious influence on epilepsy; but in combination, these two elements—bromine and potassium—are of undoubted value.[20]

While Anstie considered Reynolds's opinion as to the value of KBr, probably the strongest to have been published in England, he observed that "there has been abundant testimony from his former colleagues, and from other physicians of large experience in nervous diseases, in support of the unique value of KBr in epilepsy." [21]

Beyond this, English practitioners generally believed in the efficacy of the bromide in most cases of insomnia, false croup and asthma, colic and the night terrors of children, convulsions, and neuralgia. According to Anstie,

> there is not only a large agreement as to the positive qualities of KBr, but an agreement among the highest authorities on the respective diseases. I leave wholly unnoticed here a multitude of uses of KBr as to which there is less agreement or less weight of individual authority; for I conceive that if only thus much were fairly made out, it would be scarcely possible to exaggerate the value of the drug. . . .
>
> I cannot close this paper without quoting a remark which may tend to show, better than anything cited above, the strength of conviction with which KBr impresses some English physicians by its effects in epilepsy. I lately asked one of the shrewdest and most experienced of our authorities on nervous diseases, what he thought of the results of the introduction of KBr? He replied: "It has changed the whole prognostic significance of epileptic attacks." [22]

By 1883 there was a move both in England and America to substitute the bromide of sodium for the bromide of potassium.

T. J. Hudson, formerly senior resident medical officer at Leeds Public Dispensary, after pointing out that the "physiological actions of the salts of sodium and potassium have frequently been compared by experiment, but their relative therapeutic value has not received such careful attention," remarked:

> Bromide of sodium is now very largely used in America in preference to the potash salt, and I was induced to try it fully after observing the bad effects of the latter when given . . . for insomnia . . . and epilepsy, &c., an almost toxic effect being induced, showing itself in mental weakness, clouded intellect, failure of memory, . . . and . . . so-called "bromism," . . . The bromide of sodium lacks many of these disadvantages.[23]

Hudson's quite extensive use of sodium bromide at the Leeds Dispensary led him to the conclusion that the bromide of sodium appeared superior to the potassium "if we wish for the action of the bromine . . . only on the system, but where we desire the depressing and sedative actions to cooperate, as in pertussis, aneurism, &c., then potash salts are preferable, the alkali potash contributing by far the greater share towards these effects." [24]

Henry M. Field, professor of therapeutics at Dartmouth Medical College, took issue with authorities dating back to the early seventies, who rated the bromide of sodium inferior to that of potassium. He might, he said, have gone along with the crowd but for the fact that "every year since has the more confirmed my opinion of the superior value of the sodic salt; I have tried to observe carefully because I have to teach others as well as form conclusions for myself." [25]

> A physiological study of the two salts in question supports the following propositions:—
>
> (1.) The bromide of sodium, because it is a sodic compound, should be more congenial, less disturbing, to the fluids and solids of the body than its potassic congener.
>
> (2.) The sodium salt, in extended use, should be less depressing to the heart, all potassic salts after a time tending to produce cardiac depression.
>
> (3.) The sodic bromide is less offensive to the taste, much less irritating to the stomach.

(4.) The bromide of sodium should have equal, if not superior, general therapeutic power with the bromide of potassium, since while the former has a bromine per cent. of 78, the latter has but 66.

To which, it must be added, my clinical experience has brought me to the following conclusions:—

(1.) The bromide of sodium has equal therapeutic power, throughout the entire range of medication (with possibly an exception), with that of the bromide of potassium.

(2.) Not only this, but the bromide of sodium has *superior therapeutic value,* both from the greater mildness of its physiological impression and because of additional therapeutic applications which, were we confined to the potassium salt, would be inconvenient if not impossible.[26]

William Stephenson, of the Royal Hospital for Sick Children, Edinburgh, had firm views on the uses of sodium phosphate:

Phosphate of Soda is described in our books on therapeutics, in an almost stereotyped form, as a mild saline aperient, well adapted for females and children. Its action is such, however, only when given in large doses, and in the present day it is but rarely prescribed. . . .

The extensive field for observation afforded me at the Children's Hospital has enabled me further to investigate the matter; and in order to have as few disturbing influences as possible, the remedy was always prescribed alone, the parents being directed to put a pinch of the powder into each article of food the child received; in this way four or five grains were administered each time. . . .

The cases for which I now recommend it are chiefly the following:—

In infants who are being artificially reared, and are liable to frequent derangement of the bowels; also where the phosphatic elements in the food seem deficient, or where articles of food rich

> in phosphates, such as oatmeal, disagree; where from the character of the motions, there is a deficient or defective secretion of bile. It is thus of service in cases of chalky stools or white fluid motions. I have also found it of service in many cases of green stools. In diarrhœa, generally, it is more difficult to distinguish the class of cases. . . . It is chiefly in that class of cases which are more properly termed duodenal dyspepsia that it is of benefit. Diarrhœa after weaning is generally of this nature, and the cases are often chronic, or of some weeks' standing—the mother generally having exhausted her own and the nearest druggist's resources before applying for advice. It is also of service in some cases where the diarrhœa is due to some general cachexia. . . .
>
> In adults it has relieved constipation when taken in drachm doses in the morning. I have seen benefit also derived from it where there was a feeling of fulness and sometimes pain in the epigastrium some hours after taking food. . . .
>
> A more extended trial of the phosphate of soda in other hands than mine will, in all probability, elicit more exact information as to its action and uses. I have now given the result of two years' investigation, which . . . may contribute to redeem one more remedy from the obscurity into which it has fallen, and restore it to the catalogue of useful substances.[27]

Sea bathing has long been regarded as a healthy pursuit. Sea water is, on an average, about 2.5 percent salt (sodium chloride). In certain parts of the world—the West Indies, the shores of the Mediterranean—commercial sea salt is obtained by solar evaporation, but common (white) salt is derived from natural brine wells—such as exist in western New York State, where the salt content may run as high as 25 percent.

Salt water bathing need not be restricted to periods when the sea is comparatively warm and the air soft and balmy. A salt bath can be taken in the family bathtub, using either sea or white salt. (The latter is preferable because the former tends to be "hard," very slowly soluble in water, and more or less dirty.)

Over a dozen years ending in 1887 Henry G. Piffard, a

New York physician, observed the effects of salt water on normal and diseased human skin. He witnessed no untoward reactions in the case of normal skin. In his experience, the cutaneous affections that derived most benefit from a systematic course of sea bathing ("conducted with prudence and good judgement") were chronic eczema, sluggish psoriasis, and summer eruptions such as prickly heat, boils, and scrofulous diseases. He also found that a 5 percent brine bath, using no soap, was superior as a cleansing and deodorizing agent to soap and water in an ordinary bath. In addition, "the skin exhibits a softness and suppleness that I have never experienced from any other form of bath, be it Turkish, Russian or Roman." [28]

A more direct means of treating skin disorders externally was provided by the sulfoleate of sodium. When sulfuric acid is added to any fixed oil or fat, the oleic acid is transformed into what has been called sulfoleic or sulfoleinic acid. Add a solution of carbonate of soda under properly controlled conditions and, in twenty-four hours, sulfoleate of sodium can be separated from the mixture.

The value of sulfoleate of sodium in the treatment of skin diseases sprang from its capacity for being mixed with water, its rapid absorption by the skin, and its remarkable power of dissolving substances for which there was previously no solvent.

Its miscibility allowed it to be washed from the skin as readily as it could be smeared on. In cases where it was not desirable to keep ointment on the skin day and night, lard, Vaseline, and lanoline bases were objectionable because their removal called for soap and friction; with sulfoleate of sodium as a basis, the ointment could be removed with a little water. This same affinity for water allowed the sulfoleate of sodium to sink into the skin more readily than Vaseline or fatty substances.

> When a simple protective dressing is required for the surfaces of the skin, as, *e.g.*, in a case of acute eczema of an infant's face, vaseline or its equivalent . . . will prove of excellent service, but

> when it is desired to act upon the deeper parts of the skin or to convey medicinal substances into the blood by means of cutaneous inunction, vaseline is far inferior to lard, lanoline, oleic acid, or the sulpholeate of sodium. . . .
>
> But the chief recommendation of sulpholeate of sodium is its remarkable power of dissolving certain substances for which we have heretofore had no available solvent. Sulphur, chrysarobin, and other drugs of great value in dermatology have heretofore been applied to the skin in a finely triturated by undissolved condition. Such drugs have therefore exerted their action solely upon the surface of the skin, except when friction has forced them mechanically into the follicles, where perhaps a certain amount of absorption has taken place. The beneficial effects of these drugs thus applied can not be disputed, but when we consider that sulpholeate of sodium will dissolve at least two per cent. of sulphur or of chrysarobin, their increased efficiency can readily be imagined.[29]

On the threshold of the twentieth century G. A. Baxter of Chattanooga, Tennessee, introduced a jacket of silicate of soda for the treatment of certain spinal diseases. In presenting his body cast at a meeting of the Southern Surgical and Gynecological Association, he deplored the swing of the pendulum that had led to the almost complete abandonment of sodium silicate in favor of plaster of Paris, a loss of "much that was useful as well as ornamental in surgery; in other words, the silicate possesses so many qualities in certain cases, superior to plaster and as a substitute for metal or board splints, that it is well to recall attention to them at the present time in a presentation of the comparative claims of these two materials to our favor." While denying nothing of the usefulness and blessings of the plaster, Baxter pointed out that it had not proved satisfactory in all cases.

> Its two qualities of hardness and quick drying have enabled us to look with a great deal of leniency upon its great weight, its want

of flexibility, its proneness to rub off, to crack and to break, the difficulty had in cutting it, with the almost impossibility of readapting it, and finally the serious matter of always obtaining good plaster, so easily is it affected by changes in the atmosphere, and in consequence the danger of failure, however skilfully applied, with the added fact always in mind, that it is never suited to other than a permanent dressing, and consequently cannot be put on until swelling has subsided. These are qualities which unquestionably unfit it for many uses to which it might otherwise be excellently adapted; and when used tax greatly the skill and patience of the surgeon. There are surgeons so skilled in its use, who by their skill have been enabled to reduce these qualities to a minimum of disadvantages to themselves, but they are unquestionably ever present to the general practitioner, and afford him a great annoyance and worry. . . .

Perhaps there is no field where the benefits of plaster have been more acutely felt than in certain spinal diseases and injuries. The benefits from the use of the plaster-jacket in such cases have been so great that we have accustomed ourselves to disregard the discomfort. . . . The leather perforated jacket is good, is lighter than the plaster, and does not break down or rub off, and is ventilated by perforation, but it is costly and is not subject to readaption, and transfers, in a measure, to the instrument-maker a case that properly belongs to the doctor. The same may be said of the woven-wire jacket and all others of this nature, inclusive of the watch-spring and cotton jacket.

The lightest plaster-jacket I have known constructed for an adult weighed three and a half pounds; a lighter one than this loses the advantage of support, and is easily broken down. The jacket of baked silicate of soda, which I present to you to-day, possesses all the qualities to be found in the plaster, firmness and support, and weighs actually one pound and six ounces. It is neater in appearance and finish, can be perforated like leather for ventilation, which plaster cannot. It is even lighter than leather without its costly processes of construction, and has the same advantage over the woven-wire jacket with the additional advantage over both these latter and all others of this class, that it can be

> constructed by any surgeon at any time or in any place by the directions I shall proceed to give you.
>
> Before proceeding further let me say that the one great fault of the silicate is its slow-drying qualities; it generally takes it at least twelve, often twenty-four hours, to harden *in the ordinary way,* and that it was this quality that brought it into disrepute and made the quick-drying plaster a popular favorite with all its disadvantages. . . .

Baxter then described in some detail how he overcame this drying problem. He created a mold of the body (by first employing a rough temporary plaster cast) on which he built the silicate jacket. This was then baked. "This process of heating not only dries the silicate but bakes it as well, and renders it impervious to the action of water or the perspiration, and gives it sufficient strength to allow of it being perforated for ventilation. It is now cut from the mould with a straight incision down the centre; two pieces of leather, to which button-hooks or eyelets have been previously attached, are sewed up and down the front on each side, where the whole can be laced up solid or loosened and taken off at will." [30]

Lithium

Lithium, which was discovered in 1817 by August Arfvedson in the Swedish laboratory of Berzelius, was found to be present in the mineral waters of European and American spas used for drinking and bathing. It is possible that its use in medicine dates back to the fifth century and Caelius Aurelianus, who prescribed mineral water therapy, recommending specific alkaline springs, some of which may have contained lithium, for particular physical and mental illnesses. There are even indications that springs developed earlier by the Romans in southern and western Europe may have been used for similar purposes.

In the 1840s it was learned that lithium salts, when combined with uric acid, were able to dissolve urate deposits. Lithium was thereafter used to treat renal calculi, gravel, gout, rheumatism, and a variety of physical and mental diseases. Spas promptly exaggerated the lithium content of their springs, often adding "Lithia" to their names to attract the public.

On 10 December 1885, Esme-Félix-Alfred Vulpian (1826–1887) told the Académie de Médecine:

> In acute articular rheumatism and in acute attacks of gout, the salicylate of lithia appears to be as efficacious as the similar sodium salt. The lithia salt is occasionally useful in dissipating the still slightly painful joints left after treatment with the sodium salt.
>
> The forms of acute rheumatism in which the fibrous tissue is especially attacked, respond more readily to the lithia salt, which is also the more effective in subacute progressive articular rheumatism. By the use of the salt last named there has been obtained a noteworthy abatement of the symptoms of chronic articular rheumatism.[31]

The use of lithium in the treatment of rheumatism, gout, and like problems was largely discontinued as more effective methods were developed, but there remain enthusiasts who attribute to mineral springs the cure of a variety of ills, and in West Germany, for example, spa treatments are covered by health insurance. Spas in the Soviet Union, eastern and western Europe, the United States, Japan, and South America offer courses of treatment for a wide range of disorders—intestinal, kidney, and liver complaints, rheumatism and arthritis, anemias, gynecological problems, heart conditions, and such nervous conditions as neurasthenia, neuralgia, and "nervous breakdowns."

Today, the major professional use of lithium is in psychiatry.

PART THREE

The Twentieth Century

12

Metals in Diagnosis and Internal Medicine

By the close of the nineteenth century metals were being employed in the treatment of a variety of entities ranging from diseases of the chest, of the digestive system, and of the genitourinary organs to those of the skin and cellular tissue, and to carcinomas, extrapulmonary tuberculosis, epilepsy, and mental illness. Over better than four centuries, as one generation of doctors succeeded another, fashions and fancies in therapeutics changed, yesterday's specific becoming today's poison. Remedies once popular among doctors have in the twentieth century become obsolete, having been replaced by more specific and less toxic agents, particularly the anti-infective drugs (including the sulfonamides), antibiotics (such as penicillin), and the tetracyclines, steroids, and other synthetics.

The replacement was not of course limited to therapies deriving from metals, and the changes in the rules were not restricted to diseases long under attack. Modern science was bringing to light hitherto unsuspected, or barely suspected, problems. In fact, the twentieth-century medical researcher and practitioner is having to face four types of situation: First, there are entities for which cure or control measures exist and which would be totally eradicated if they were not neglected in many parts of the world—including pockets of neglect in the medically advanced countries; such entities include smallpox, yellow fever, malaria, various other fevers, infectious and contagious diseases, venereal diseases, and more recently poliomyelitis.[1] Secondly, there are entities that were passed down with little or no solution in sight—degenerative diseases of the vital organs, carcinomas, the rheumatoid diseases, hepatitis, epilepsy, mental illness, and those hardy perennials, influenza and the common cold. Thirdly, there are newly recognized entities, including deficiency and excess diseases, congenital defects, and blood dyscrasias. Finally, there are the inevitable results of accident—trauma and burns.

Metal Deficiency and Excess

Metabolism is the process by which all foods, fluids, gases, and other substances taken into the body are utilized by the body in its general maintenance, survival, growth, repair, and function. When diet is improper—either because good food is scarce, or because the individual cannot afford it, or because the individual *can* afford to indulge his whims—deficiency disabilities are likely to result. Among the earliest deficiency diseases to be recognized were rickets and scurvy. Rickets, a vitamin D-calcium-phosphorous deficiency, may have been recognized as early as the days of Homer (c. 850 B.C.); scurvy, a deficiency of vitamin C arising from a lack of fresh fruit and vegetables, appeared in 1218. But there is another road to deficiency. The human body contains a number of trace metals, the retention of which is essential. Their loss can lead to problems; so can an oversupply.

> As more and more about the physiologic roles of minute amounts of metallic elements comes to light, a rising potential is indicated for advances in human health—in growth and development research, in diagnosis, and in clinical specialties all the way from prenatal to geriatric medicine. But leaders both in nutrition and in clinical research say the quest has only begun.
>
> New knowledge on the presence and body sites of various trace elements is helping to uncover the ways each metal may interact at the molecular level, with proteins, nucleic acids, or hormones; and to determine whether the trace particle acts as a catalyst, as part of a molecule, or both. In a majority of the functions already known, the trace metal acts as a catalyst, influencing enzyme production. In most trace minerals, either a deficiency or an excess may precipitate disorder; in some, the span between too little and too much is a narrow one.[2]

Zinc deficiency was, until quite recently, thought to be impossible in man because of zinc's liberal presence in his food and water intake, but in the late 1950s and early 1960s zinc deficiency was detected in malnourished populations. Ananda S. Prasad, now a professor at Wayne State University School of Medicine (Detroit), and James A. Halsted, who was at the time Fulbright Professor of Medicine at the Shiraz (Iran) Medical School, studied a number of Iranian and Egyptian males who were suffering from iron deficiency anemia, sexual underdevelopment, and dwarfism. The first patient examined by Prasad, in the fall of 1958, was twenty-one but "looked like a ten-year-old boy and had severe anemia. . . . The nutritional history was interesting in that this patient ate only bread made of wheat flour, and the intake of animal protein was negligible. He consumed nearly one pound of clay daily." [3] Subjects generally were nineteen or twenty, had an average bone age of around ten years, and were 20 percent shorter and 45 percent lighter than normal.

The administration of oral iron eliminated the anemia, but subnormal concentrations of zinc in plasma, red cells, urine, sweat, and hair remained to be dealt with. Zinc sulfate capsules given to several dwarfs resulted in growth of external genitalia, the appearance of pubic, axillary, and facial hair, and significant increases in height.

Prasad concluded that there "are several causes of zinc deficiency in human subjects in the Middle East: (1) unavailability of zinc from cereal diets normally consumed, (2) excessive blood loss due to hookworm infestation (seen in Egypt), and (3) loss of zinc by sweating in hot tropical climates." [4]

In 1968 it was determined that patients with chronic venous leg ulcers and patients with bedsores have a significantly lowered mean plasma zinc concentration and that geriatric patients treated with oral zinc sulfate have fewer bedsores. In the early 1970s R. J. Clayton of the Royal Free Hospital, London, undertook a double-blind trial of oral zinc sulfate in patients with leg ulcers but reported that the "evidence that oral zinc enhances the healing of leg ulcers is inconclusive." [5] In 1972 T. Hallböök and E. Lanner of the University Hospital of Lund, Sweden, reported on the effectiveness of oral zinc sulfate in the healing of venous leg ulcers.[6] In commenting upon this report, Daniel B. Stone of the University of Nebraska Medical Center questioned the suggestion of the authors "that patients respond best to treatment with zinc when they have low blood zinc concentration. My reservation is that blood zinc concentration does not accurately indicate the concentration of zinc inside cells." Stone went on to assess the current status of zinc in treatment:

> Zinc deficiency crosses my mind in patients with varicose ulcers, in patients with slow wound healing, and after operations on the larynx or trachea. I think about zinc especially in patients who

are prone to zinc deficiency: liver disease; diabetes mellitus; renal disease; and in all patients who have not been eating well.[7]

Concurrently Aba Marshak of Haifa, Israel, and Gabriel Marshak of the University of Cincinnati Medical Center, Ohio, were finding oral zinc sulfate therapy useful in the treatment of granular tumors of the vocal cords forming subsequent to the performance of a tracheotomy or otherwise. They found it especially of value because it is simple to administer, preserves the anatomical and functional integrity of the vocal cords, offers quick symptomatic relief, avoids surgical incision, and, at the right dosage, is not toxic.[8]

All human body tissues contain some zinc with concentration high in blood, bone, kidney, liver, and certain muscles and highest in the prostate gland and the inner coating (choroid) of the eyes.

Under normal conditions, zinc concentrations remain remarkably constant in the plasma and tissue of the body despite a rapid turnover of the metal. But acute alterations in zinc metabolism occur under stress. Since zinc is not stored to any extent in the body, even with an adequate dietary intake, disease or physical injury can, by increasing zinc excretion, induce tissue depletion. The amount of zinc in the urine increases in liver disease, diabetes, some kidney problems, and following the taking of certain drugs. Zinc plays an important role during the healing of wounds and burns, and following surgery. The need for the therapeutic replacement of zinc is on the increase even in the more affluent nations.

Zinc deficiency in plants has been found in at least 30 states. Dr. Frank G. Viets, Jr. of the U.S. Department of Agriculture says crop zinc deficits have increased spectacularly over the past 20 years. One troubled prediction is that unless leached-out soil zinc is replaced on a wide scale, more and more people will develop symptoms of zinc and other trace metal deficiencies.[9]

But zinc must not be ingested recklessly. An excess will produce fever and gastrointestinal symptoms, and zinc oxide fumes can bring on pneumonia, with fever, malaise, and mental depression. Chronic zinc toxicity may lead to a copper deficiency.

Copper deficiency in man is extremely rare because dietary copper is so abundant. In fact it is difficult to prepare an acceptable diet that contains less than 2 milligrams of copper daily (ordinarily the body will retain through absorption 0.6 to 1.6 mg.) and content may run as high as 5 or even 10 milligrams if the diet includes such copper-rich foods as oysters, liver, mushrooms, nuts, and chocolate. Copper serves in brain and blood function and in the production of red cells. When a deficiency occurs, it is the result of limited intake (as in the case of infants fed only milk, which contains little or no copper), malabsorption, or excessive loss of the metal (due, for example, to diarrhea). Malnourished infants are targets for anemia, neutropenia (decrease in certain white cells), and bone lesions, for all of which cure involves the administration of copper. In Africa, Asia, the Middle East, and Central and South America, children from one to four are subject to kwashiorkor (the name deriving from an African dialect word meaning "red boy," given because the disease is accompanied by depigmentation), a protein deficiency disease also involving copper deficiency. Albinism has been attributed to a congenital copper deficiency.

In 1962 a group of residents from the departments of neurology, pediatric neurology, neuropathology, and dermatology, headed by John H. Menkes of the University of California at Los Angeles School of Medicine, described

> what appeared to be a new degenerative disease of the central nervous system. The condition was transmitted as a sex-linked recessive trait. Affected male infants were noted to have peculiar

> stubby, white hair, early and severe physical and mental retardation, and widespread focal cerebral and cerebellar degeneration. While the marked inanition found in many patients even prior to the onset of neurological symptoms pointed to a generalized metabolic disorder, we were unable to go any further in understanding the disease.[10]

In 1966 it was named kinky-hair disease by John S. O'Brien and E. Lois Sampson of the Division of Chemical Pathology at the University of Southern California School of Medicine.

Between 1962 and 1972 there were only five reported cases of the disease and it came to be considered very rare, but when David M. Danks and a group of workers from the Royal Children's Hospital Research Foundation in Melbourne (Australia) detected seven cases in five families over a three-year period they concluded it was not so rare. Their studies revealed "a defect in the intestinal absorption of copper in Menkes' kinky-hair disease. . . . The copper deficiency which follows the impaired copper absorption in these patients provides a satisfactory explanation of the known features of the disease." [11]

> These findings open the possibility of therapeutic intervention to prevent the tragic progression of the brain damage. Unfortunately, severe damage has generally occurred before clinical diagnosis is possible. Effective treatment will need to start before symptoms develop, and may therefore be possible only in new cases in infants born into families known to be at risk. In addition, there is some reason to fear that copper deficiency is already established before birth (perhaps because of a defect in placental transfer). The liver of a normal newborn child contains large stores of copper, and copper deficiency does not develop during many months of low copper intake, even in premature babies. However, copper depletion may develop more rapidly in babies with Menkes's syndrome if the large amount of copper normally excreted in the bile cannot be reabsorbed. Ideal treatment would involve finding an oral preparation of copper which could bypass

> the absorptive block, but this may prove impossible and parenteral administration may be necessary. Copper-albumin or copper-histidine complexes can be given intravenously, but there is no experience with intramuscular preparations.[12]

Situations involving copper excess arise far more frequently.

> Individuals have been acutely poisoned by ingesting grams of copper salts—usually the sulfate ("bluestone")—or by having had solutions of copper applied to large areas of burnt skin. Most of the clinical experience of this form of toxicity comes from India where the ingestion of copper sulfate is not infrequently a means of committing suicide. . . . There is only fragmentary information on the mechanisms by which acute copper toxicity is effected in animals or man. . . .
>
> It is the natural, chronic copper poisoning of man, known as Wilson's disease, which is of greatest relevance since its study has afforded considerable understanding of human copper metabolism. This disorder, inherited in autosomal recessive fashion, occurs in perhaps one in 200,000 individuals around the world, except for possibly an even lower incidence among pure Negroes. It is the only form of chronic copper poisoning known to occur in man. Indeed, in contrast to animals, no diet or exposure to copper-rich dust seems able to induce copper toxicity in man except in patients with Wilson's disease. Even in copper miners, who may breathe and eat meals in an atmosphere nearly opaque with 1 or 2 percent copper ore, concentrations of copper in serum and liver are normal and evidence of toxicity is absent.[13]

In Wilson's disease, liver copper may rise to twenty times its normal level while ceruloplasmin, a copper-binding plasma protein, drops to an abnormally low level. Ultimately the excessive liver copper pours into the circulation, apparently destroying red blood cells. Released more slowly, it seems to invade all body tissues, including the corneas of the eye and the kidneys (where it does not appear to do much damage) and the central nervous system (where the invasion may prove significant).

Carl C. Pfeiffer of the New Jersey Neuropsychiatric Institute in Princeton has attributed excess copper blood levels in schizophrenics at least in part to excessive copper content of drinking water. "Many American families pump their own household water from the soil. In shale soils the water may be high in sulfur dioxide (sulfurous acid), and in marshy soils excess carbonic acid may be present. Either or both of these acids will remove copper from the pipes to produce water as high as 5 ppm of copper. . . . The United States Public Health Service rules that water with more than 1 ppm of copper is unfit for drinking purposes." [14] I. Herbert Scheinberg and Irmin Sternlieb of the Albert Einstein College of Medicine disagree. "It seems, incidentally, worth noting that neither acute nor chronic copper toxicity has been associated with the use of water pipes or cookware made with copper." [15]

Iron, of all the metals in the body, has been studied most religiously, with investigations dating back to Sydenham, Menghini, and Berzelius. These explorers were aware of a relationship between iron and the blood, but they certainly did not know that approximately 65 percent of bodily iron is to be found in hemoglobin (where its primary function is the transporting of oxygen), nor did they think in terms of iron deficiency. The normal adult ingests 10 to 20 milligrams of iron daily, but the amount actually absorbed depends on individual requirements determined by a variety of factors. By way of examples, the female while menstruating loses about twice as much iron daily as she (and the male) would normally lose, and the pregnant female needs additional iron to replace roughly 300 milligrams stored in the fetus during gestation. Hence, iron deficiency anemia occurs more frequently in women (and children) than in men. Iron deficiency also results from severe infections, an absence of free hydrochloric acid in the stomach, chronic diarrhea, chronic bleeding (as from hemorrhoids, menorrhagia, and the like), and impaired absorption. Iron deficiency produces such symptoms as lack of appetite, dizziness, and proneness to fatigue. While the body appears to have some capacity for controlling the amount of iron absorbed, the deposit of excessive amounts of iron in tissues is possible. If no tissue damage results, the overloading is considered secondary and is known as hemosiderosis; if tissue damage is involved, the generalized overloading is called hemochromatosis. The latter is characterized by a yellowing or bronzing of the skin and may invite cirrhosis of the liver, fibrosis of the pancreas leading to diabetes mellitus, and congestive heart failure.

Iron deficiency problems are met by myriad iron preparations, each boasting a particular advantage over the others. The drugs of choice would seem to be ferrous sulfate (usually given in tablet form to adults and as a syrup to children), ferrous gluconate, and iron dextran injected intramuscularly. Hemochromatosis is best controlled by bloodletting.

Pica, from a Latin word meaning magpie, involves a craving for such substances as starch, clay, ashes, plaster. Why do some people eat dirt or starch, or chew on ice at all hours of the day? In studies undertaken over the past fifteen years, most investigators support a connection between pica and iron deficiency. The relationship of pagophagia (ice eating) to iron deficiency was the most marked, followed by geophagia (dirt or clay eating) and amylophagia (starch eating). Sometimes iron deficiency anemia was the cause of pica; sometimes the reverse was true.[16]

Manganese as a trace metal enjoys a somewhat unusual position. Manganese deficiency, while operative in animals (causing muscular incoordination, the weakening and bowing of bones, and reproductive defects), does not apparently occur in man. On the other hand, manganese toxicity is seen in man (following chronic inhalation of manganese dust by workers) but not in animals.

> The insidious development, after months to years of exposure, of psychiatric abnormalities and a neurological disease resembling parkinsonism or Wilson's disease characterizes the condition. Surprisingly, in the light of our knowledge of the latter disease, no increase of the manganese concentration of any tissue except lung accompanies the syndrome, although the appearance of manganese in the urine may be of diagnostic value. Though incapacitating because of weakness, spasticity, tremors, and disturbances in gait, the disease does not appear to be fatal.[17]

Chromium is a newcomer among the biological metals known to be essential to man. In 1955 a missing factor in diet became known as the "glucose tolerance factor"; in 1959 the factor was recognized as chromium deficiency. Most elderly people in the United States have impaired glucose tolerance, and 50 percent of those treated with trivalent chromium have been restored to normal glucose tolerance. But the fact remains, as pointed out in 1972 by D. K. Michael Hambidge of the University of Colorado Medical Center, that by and large the senior citizens of this country show lower concentrations of chromium than adults in the Far East, Middle East, and Africa.

Thus chromium deficiency may be attributed to two factors: poor dietary habits; food-processing methods that, through overheating, remove chromium or convert it into a form that is not readily absorbed. (Sugar refining and the milling of wheat remove a large percentage of chromium.) Foods with a high biologically active chromium content include brewer's yeast (beer contains more chromium than milk), black pepper, thyme, liver, beef, mushrooms, parsnips, *cooked* tomatoes, corn oil, and bread. The tissue chromium drop-off that occurs in babies may be attributable to the extremely low chromium content of cow's milk and many infant formulas.

> The elderly [and the very young] are not the only group affected by apparent chromium deficiency; both insulin-dependent juvenile diabetics and maturity-onset diabetics also are at risk. The insulin-requiring patients may absorb but underutilize chromium, as shown by excessive urinary excretion; and, whether as a cause or an effect, abnormal chromium metabolism seems to accompany the disease.[18]

Kwashiorkor is usually accompanied by a deficiency of sugar in the blood and impairment of glucose tolerance, and Walter Mertz of the USDA's Human Nutrition Research Division, a pioneer in the study of chromium metabolism, has raised the question whether repeated pregnancies may not cause progressive impairment of glucose tolerance in women.

Knowledge of chromium metabolism is at a stage where there are more questions than answers. But some things are certain. Chromium does not act pharmacologically to reverse a sugar deficiency in the blood; therefore an improved glucose tolerance following chromium dosage indicates that a chromium deficiency was present.

Cobalt is an essential component of vitamin B12, but it functions in the body of man in forms other than as part of vitamin B12.

While an excess of cobalt leads to an oversupply of red blood cells (polycythemia), cobalt has been successfully employed to increase the red blood count in nutritional and secondary anemias. In states of chronic inflammation, cobalt can promote the absorption and utilization of iron for hemoglobin synthesis. Aside from its red blood cell action, excess cobalt has been known to cause depression of muscular coordination, loss of appetite, and goiter.

An interesting epidemic, attributable to cobalt, hit the Quebec area in 1965. It involved muscular disease of the heart and resulted in death in half the cases. Since the outbreak was localized, a toxic cause seemed probable. It was found that all the patients were heavy drinkers of a particular brand of beer.

Before 1965 it had been shown in Scandinavia that cobalt "stabilizes" the froth on the top of beer and various breweries started adding it to their beer. "With the advent of newer, more powerful detergents and the inadequate rinsing of beer glasses, some of the detergent remained when the glass was next used for drinking beer. The film of remaining detergent was sufficient to impair the froth of the beer. The addition of cobalt chloride to the beer 'stabilized' the froth despite the film of detergent." [19] Subsequently cases of cardiomyopathy were reported from other areas where cobalt salts were added to beer.

Magnesium, found in many foods including meat, cereals, vegetables, and milk, is an essential component of chlorophyll, where its structural role is somewhat similar to that of iron in hemoglobin. Normal adults ingest about 300 milligrams a day, of which approximately one-third is absorbed through the small intestine. The kidneys serve as the final regulator of magnesium balance. Sweating causes a loss of magnesium through the skin.

Knowledge of the function of magnesium in the human body is still limited. It is certain that it is an activator of many enzyme systems, and if it is lacking or fails to function in these systems, physiological disturbances may result. Following a major, long-term depletion of total body magnesium content, the central nervous system, the peripheral nerves, and the myoneural junctions become involved. Magnesium injected intravenously depresses the entire central nervous system (resulting in a lowering of blood pressure). The effect on the central nervous system is used to combat the convulsions of tetanus and convulsions associated with labor. Magnesium excess occurs during kidney failure, certain disorders of the endocrine glands, an artificial lowering of body temperature, or when too much magnesium is administered as a drug.

The history of **selenium** in human metabolism has been interrelated with vitamin E (discovered some fifty years ago). Researchers were aware that vitamin E was involved with a mysterious "factor 3." In 1957 Klaus Schwartz, now at the National Institute of Arthritis and Metabolic Diseases, Bethesda, Maryland, pinpointed factor 3 as selenium. But high toxicity has limited investigation of its action in the human body. The nutritional requirement of selenium in the diet is 0.1 to 0.3 parts per million. Selenium in the range of 2 to 10 ppm produces chronic toxicity, over 10 ppm, sudden death. This does not leave much room for experimentation.

Donald F. L. Money, director of the Animal Health Laboratory at Wallaceville Animal Research Center (New Zealand) has assembled telling evidence that the human sudden infant death syndrome is strongly related to selenium or vitamin E deficiency. Crib deaths, which in the United States run as high as 35,000 a year, occur after the first and mostly during the second and third months of life. Money has pointed out that practically no breast-fed infants die in this manner. Human milk contains up to six times as much selenium (and about twice as much vitamin E) as does cow's milk.

In recent years the question has been raised whether selenium inhibits or incites cancer. A nineteen-city survey that set human blood selenium content against cancer deaths per 100,000 showed Rapid City, South Dakota, with the *highest* blood content (0.256 ppm) and the *lowest* deaths (94 per 100,000). Conversely, in Lima, Ohio, with the lowest selenium blood level (0.157 ppm), cancer deaths ran 188 per 100,000.

> Many factors other than selenium content of the blood undoubtedly figured in the cancer mortality in these various cities. Thus the correlation cannot be said to show that higher blood selenium levels *prevent* cancer. It is possible, however, to use these data as evidence that chronic blood selenium values at levels of approximately 0.25 ppm . . . *do not increase the incidence of cancer* compared to populations where the selenium of the blood is 0.15 ppm.[20]

Selenium has recently been found to counteract the toxic effects of mercury in food.

Molybdenum deficiency contributes to dental caries in humans. This conclusion was reached more through wondering why the incidence of caries was lower in certain places than investigating a prevalence of caries in others. In parts of Hungary, for example, less caries occurred than might have been expected from the fluorine content of the water supply; it was found that the drinking water was high in molybdenum. The cities of Hastings and Napier in New Zealand shared a water supply, yet there was markedly less caries in Napier; there was, however, a considerably higher concentration of molybdenum in the soil in which the citizens of Napier grew their vegetables. In Somerset, England, the incidence of caries is high in children from areas where cattle are subject to molybdenum deficiency. Molybdenum has been shown to reduce the solubility of teeth in acid and to reduce the acid output of the salivary glands. Molybdenum deficiency in the soil has been held responsible for cancer of the esophagus in the Bantus of southern Africa.

The normal ranges of serum **nickel** in man appear to be constant from birth to old age, not even being significantly altered by pregnancy. While there is a below-normal concentration of nickel in hepatic cirrhosis, investigations to date have tended only to reveal situations in which nickel levels have become elevated. It has been known since 1963 that serum nickel levels rise abruptly within the first twenty-four to thirty-six hours after myocardial infarction—though the like does not occur in patients with acute myocardial ischemia (temporary inadequacy in the blood flow) without infarction. Heightened nickel levels have been found in the blood and neoplastic tissues of cancer patients and in the blood of leukemia patients. Increased nickel has been found in the blood and liver of stomach cancer patients.

Aluminum in serum has been found to accompany acute pulmonary infarction. When a patient is treated for long periods with aluminum hydroxide in chronic renal failure, small quantities of aluminum are deposited in the bones. Concentrations of aluminum, **silicon,** and **titanium** in the blood are depressed in patients with diabetes mellitus. Patients with hepatitis show increased concentrations of **gallium** in the blood—the more severe the illness the higher the concentration. Still higher values are found during the early stages of recovery. **Cadmium** concentrations in blood serum, liver, and kidneys are increased in patients with bronchial cancer.

Because it could not be found in newborn infants, **tin** was not regarded until recently as an essential trace element, but it has now been established that one or two parts per million of tin are necessary in the diet. The fact that **vanadium** deficiency has an adverse effect in animals cannot be used to anticipate the same results in man, but, says Leon L. Hopkins, Jr. of the U.S. Department of Agriculture, "if man continues to refine and purify his diet without consideration for replenishing extracted trace elements such as vanadium, problems that are speculation today may prove very real tomorrow." [21] **Strontium** is also in line for being recognized as an essential trace metal.

The designation "trace elements" arose from the fact that their presence was known or suspected but their concentrations in the body were believed to be unmeasurable; today the content can be measured with precision, but the name has persisted. There are, however, three metals that are present in the body in too great an amount to justify their being classified as trace elements, which nonetheless present deficiency and excess problems. They are sodium, potassium, and calcium.

A deficit of sodium is practically never due to a low sodium intake because sodium is effectively conserved by the normal kidney. Excessive loss occurs as the result of kidney abnormality, from the bowel, or through the skin. Kidney loss may also result from drugs and from Addison's disease (adrenal cortical insufficiency). Bowel loss results from diarrhea and drainage from fistulas of the bowel, the pancreas, and the biliary ducts. Skin loss involves heavy sweating, burns, and fibrocystic disease (in which the sweat carries an unusually high sodium concentration). Deficiency conditions are usually met with large doses of sodium chloride. (Sodium has been found to be the least toxic of the body metals.) Sodium deficiency is a contributing factor in heat prostration—unacclimated individuals in hot dry environments should increase their dietary salt through the daily use of sodium chloride tablets or sodium chloride in water. An excess of sodium may be expected in edema. The congestive heart failure patient has a limited ability to handle sodium (just as a diabetic is limited in respect of dextrose). Such excess is controlled through dietary restrictions or administration of diuretics.

Differentiation between potassium excess and potassium deficiency is not easy. Rapid and slowed heart action may arise from both, although the latter is more often associated with potassium intoxication. Furthermore, cardiac function, smooth muscle function, and skeletal muscle function are related to the rate of percentage change in potassium concentration rather than the maintenance of any absolute level.

Depletion of body potassium produces a clinical picture involving marked weakness of muscles, rapidity of heart action, rapid shallow respiration, obstruction of the small intestine, and sometimes subnormal blood pressure and even paralysis. A variety of factors may produce deficiency. They include excessive vomiting, diarrhea, drainage of potassium-rich gastrointestinal secretions, diabetic acidosis (wherein cellular potassium is lost through the urine), poor nutritional intake, chronic disease states, and certain forms of kidney inflammation (under normal conditions potassium leaves the body almost entirely by renal excretion). Patients treated with potent diuretics are particularly prone to potassium depletion. By way of treatment, oral or intravenous administration of potassium has become commonplace. In such administration care must be taken to see that the rise of serum potassium is not too rapid or cardiac standstill may result. However, if the patient does not have renal disease and the urinary output is adequate, potassium therapy may generally be safely conducted.

In treating starved and dehydrated children, the administration of large amounts of sodium in an attempt to correct acidosis may lead to potassium deficiency and sodium intoxication. To avoid this, unless contraindicated by inadequate kidney function or existing potassium intoxication, potassium should be given in amounts equivalent to the sodium being administered.

Patients receiving cortisone and like steroids or thiazide diuretics may develop potassium deficiency as a result of increased renal loss. Potassium chloride may be administered concurrently as a preventive. Low levels of blood potassium increase the sensitivity of the heart to digitalis. Irregular heart action developing from this situation may be eliminated with potassium. Potassium therapy must be employed cautiously in patients with congestive heart failure because of an accompanying delay in renal excretion. Patients emerging from diabetic coma invariably have a serious potassium deficiency (generally associated with phosphate deficiency). Treatment involves intravenous administration of a solution of potassium phosphate following several hours of intravenous infusion of insulin and glucose. Minor potassium deficiencies may be overcome through a diet based on fruit juices, meat broths, and other foods with high potassium concentration.

Normal individuals under physical stress, such as that incurred during intense prolonged participation in contact sports, may be subject to acute muscle damage. A lesser degree of exercise will produce cramps and even tissue death in patients with preexistent muscle problems, hereditary or acquired. Skeletal-muscle blood flow is normally very low at rest but rises sharply during muscular work. Among the many factors in muscle metabolism, potassium seems to play a major role; the intensity of muscular work, the release of potassium, and the rise in muscle blood flow are closely interrelated. Consequently, if there were a potassium deficiency, the likelihood of muscular damage is enhanced. For example, when military recruits and football players are trained in hot climates, there is a tendency toward overheating and even heatstroke with an accompanying potassium loss. The immediate answer would seem to be potassium supplementation, but there is a danger of developing excess potassium if a supplement is given in a single large dose just before exercise. It has been suggested that the training-loss potassium be made up by smaller doses taken at mealtimes, but "possible risk has not been evaluated and, in consequence, [it] cannot be recommended without reservation. Perhaps the safe solution lies in less intense activity during the first few weeks of training in the heat, which in turn would not lead to comparably severe potassium deficiency." [22]

Elevation of serum potassium may result from inadequate

excretion due to reduced kidney function, from a transfer of cellular potassium to extracellular fluid, from contraction of volume of extracellular fluid, or from injudicious use of potassium salts. Adrenal insufficiency may produce minor increases in serum potassium. As has already been noted, symptoms found in hyperpotassemia resemble those of hypopotassemia. They include loss of strength, hypotension, mental confusion, abnormal sensation (such as numbness), pallor, and slow and irregular heart action. Hyperpotassemia may be treated by intravenous infusion of sodium bicarbonate or by intravenous injection of calcium chloride, calcium gluconate, or dextrose. When abnormalities subside, excessive potassium in the body may be removed by administering sodium polystyrene sulfonate orally or rectally as a retention enema.

Calcium accounts for almost two-thirds of the dry weight of human bone. Ninety-nine percent of the calcium in the body is in the skeleton and teeth. The remaining 1 percent is in the blood, in intracellular fluid, and in various soft tissues. In addition, about 1 percent of skeletal calcium is freely interchangeable with extracellular fluid. Despite the smallness of the amount of calcium not in bone and teeth, calcium is essential to normal function of the heart (where it activates myocardial contraction), to coagulation of the blood, to the functioning of neuromuscular tissues, to the secretion of hormones, and for lactation. It is also important in the acid-base equilibrium of the tissues, proper proportions of sodium, potassium, and calcium being necessary. Bone calcium is not stable. From fetal life to senility bone is being dissolved and reconstituted with varying rapidity.

Quantitative relationships between absorption and excretion of calcium are difficult to establish because of variables in the amount of excretion in the feces. The normal human dietary intake ranges from 0.5 to 1.0 grams per day, of which—depending on age, general state of calcium balance, previous dietary history, and actual dietary intake of vitamin D (of which an adequate supply is essential to calcium absorption)—from 15 to 70 percent is absorbed. Milk contains 1.4 grams of calcium per liter, and cheese, green vegetables, and egg yolk are high in calcium, but their high phosphorous content tends to interfere with calcium absorption. (Where milk and milk products are restricted in the diet, dibasic calcium phosphate is used as a supplementary source.)

A deficiency of calcium in the blood serum may result in tetany. This nervous affection is characterized by irritability of all muscles, leading to intermittent spasms, particularly of the extremities, but also affecting smooth muscle in the iris, bronchi, stomach, intestines, and urinary tract. When spasm of the larynx occurs, it may be severe enough to result in suffocation. Irregular heart action may also result.

Treatment involves the intravenous administration of a soluble calcium salt (usually calcium gluconate) followed by oral administration of calcium lactate. When hypocalcemia is due to vitamin D deficiency, the acute symptoms are controlled with an intravenous calcium injection before vitamin D is administered.

An excess of calcium in the blood serum leads to conditions more severe than tetany and the convulsions associated with it. A high content of calcium blocks the transmission of nerve impulses. Weakness, muscle and joint pain, loss of appetite, loss of weight, impaired kidney function, constipation, and various irregularities of the heart have been known to follow. Hyperparathyroidism (a condition caused by increased parathyroid secretions) results in an elevation of serum calcium concentration, increased excretion of calcium, and demineralization of bone. Nature's attempt to counteract this weakening results in a disturbance in the calcium-phosphorous metabolism and to a condition known as *osteitis fibrosa cystica*. If, however, calcium intake is sufficient to balance the increased excretion, no bone change will occur. Instead one may look for the formation of kidney stones. Very large doses of vitamin D can cause hypercalcemia, as can the administration of sex steroids to women suffering from spreading breast cancers. Among drugs useful in the treatment of hypercalcemia are sodium or potassium phosphate, sodium sulfate, cortisone, mithramycin, and disodium edetate. Disodium edetate also makes insoluble forms of calcium soluble and hastens excretion. It has been used to dissolve kidney stones and abnormal calcium deposits, and it has been suggested that this action might be useful in the treatment of sclerosed coronary arteries.

In increasing age the normal replacement of wasted bone is diminished and bone tends to become more porous, a condition known as osteoporosis. The osteoporotic individual is liable to spontaneous fractures. The fundamental defect is a deficiency of osteoid tissue, the protein matrix of bone. Even with a normal calcium metabolism, a matrix formation deficiency leads to loss of bone and loss of calcium from the body, because osteoporosis is primarily a disorder of protein metabolism and only secondarily related to calcium and phosphorous metabolism. Estrogenic and androgenic hormones have been known to relieve pain associated with osteoporosis, but they do not cure, and optimum rest on a firm bed and balanced activity is the only prescription that can be offered.

Metallic Medicaments

It would be cumbersome, tedious, and inappropriate to catalog the myriad uses of metal salts and combinations in modern medicine. Discussion must be limited to those used in combatting major diseases or in therapies that are unusual or outstanding in their effectiveness.

It has been common practice in cases of congestive heart failure to use diuretics in addition to digitalis at the outset of treatment. The objective is to draw off, through increased urine secretion, excess fluids (especially in left-side heart failure) and prevent the development of pulmonary edema. Among available diuretics the mercurials, once described as "the greatest advance in the treatment of congestive failure since Withering," are among the most potent and most dependable, but they are usually injectibles and have given ground to more modern, oral diuretics which exhibit less toxicity. The half-dozen mercurials now on the market differ little from one another.

Sodium nitroprusside administered intravenously promptly reduces blood pressure, but the infusion must be continuous because blood pressure returns to hypertensive levels as soon as the infusion is stopped. Sodium nitroprusside is consistently effective even when hypertension is resistant to other agents and would almost invariably be the drug of choice save for the fact that its administration calls for close supervision by trained personnel.

In 1869 Paul Langerhans, a young German medical student, discovered the islands in the pancreas that were to bear his name. These islands secreted a hormone which would later become known as insulin, the antidiabetic hormone. Insulin is readily crystallized as a zinc salt. (Nickel, cadmium, and cobalt are also effective.)

Obese patients with mild uncomplicated diabetes may control their disease by means of a low-caloric diet—without insulin; other patients, not obese, whose diabetes is mild and began during their adult years, can control it by diet plus an oral hypoglycemic drug; but all diabetic children, all diabetic adults who have lost an excessive amount of weight, all diabetics with acute complications, and all individuals with severe diabetes need regular hypodermic injections of insulin.

There are three categories of insulin preparations: rapid-acting, intermediate, and slow-acting (or long-acting).

Rapid-acting insulin is indispensible in the treatment of diabetic coma and acidosis resulting from inadequate insulin dosage. Emergency administration is intravenous and only soluble zinc crystals may be used. Also acting within one hour (but reaching a delayed peak) is prompt insulin zinc suspension (or semilente insulin) administered subcutaneously—never intravenously.

The intermediates include globin zinc insulin—a sterile solution of insulin modified by the addition of zinc chloride and globin (obtained from beef blood), lente insulin (a combination of 30 percent semilente and 70 percent ultralente insulin), isopane insulin suspension which contains insulin, protamine, and zinc, and protamine zinc—a sterile suspension of insulin modified by the addition of zinc chloride and protamine.

Ultralente insulin (extended insulin zinc suspension) is the longest acting, with onset occurring in four to six hours, peak effect at sixteen to twenty-four hours, and a duration running to thirty-six hours.

The cause of rheumatoid arthritis is unknown. The disease primarily affects the joint cartilages and bones, destroying the joint cavity, which becomes filled with adhesions. Atrophic changes appear early, involving the soft tissues about the affected joints.

The use of gold as a therapeutic agent dates back to antiquity; the use of gold salts in the treatment of rheumatoid arthritis was fortuitous. Early in this century rheumatoid arthritis was believed, at least by some, to be an effect of the tubercle bacilli on the joints, and gold cyanide had been demonstrated to inhibit the tubercle bacillus. By 1929 gold had been found beneficial in the treatment of rheumatoid arthritis.

Preparations in common use today include such organic salts as gold sodium thiomalate and gold thioglucose and such inorganic salts as gold sodium thiosulfate. They are injected intramuscularly, are absorbed slowly, and circulate in the protein fraction of the plasma. The therapeutic effectiveness of gold salts has been debated and there has been no general agreement as to whether they should be employed at the earliest possible time or only after other measures have failed. Their supporters concede that they are of value in the treatment of rheumatoid arthritis only when the peripheral joints are involved. Recent investigations suggest, however, that gold thiomalate is useful in the treatment of patients with rheumatoid arthritis involving a marked decrease in white corpuscles (leukopenia) but not splenomegaly (enlargement of the spleen) which had been thought to cause leukopenia.[23]

Clearly gold salts are not the final word in the treatment of rheumatoid arthritis but they have so far withstood the test of time.

No one is known to have estimated how high a mountain and how wide a lake would be formed by the tablet and liquid antacids consumed daily by digestive-tract-conscious Americans. However, in reporting the generally nontoxic effect of aluminum hydroxide in the treatment of chronic renal failure, John C. Krantz, Jr., director of pharmacologic research at the Maryland Psychiatric Research Center, hails the information as "glad tidings to the great number of patients who ingest the multiplicity of antacids that contain aluminum salts such as aluminum hydroxide." [24]

Antacids are far from restricted to aluminum salts. They include precipitated calcium carbonate, calcium hydroxide, tribasic calcium phosphate, calcium tartrate, magnesium carbonate, tribasic magnesium phosphate, magnesium hydroxide, magnesium oxide, magnesium tartrate, potassium bicarbonate, and sodium bicarbonate. Various commercial preparations combine two or more antacids—calcium and magnesium carbonates, calcium and magnesium tartrates, magnesium and aluminum hydroxides, magnesium trisilicate with aluminum hydroxide—but these combinations are apparently not superior to calcium or magnesium carbonate in neutralizing action and are largely window dressing.

Aluminum hydroxide gel, aluminum phosphate gel, and precipitated calcium carbonate are the antacids most commonly used in the treatment of peptic ulcers, the aluminum phosphate gel being particularly recommended for peptic ulcer patients who have a deficiency of pancreatic juice, a chronic diarrhea, or a dietary deficiency of phosphate.

Bismuth subcarbonate is used in the treatment of such gastrointestinal disturbances as gastritis, enteritis, diarrhea, dysentery, ulcerative colitis, and duodenal ulcer.

In cirrhosis of the liver, fluid retention leading to edema springs from a variety of causes. Diuretics and restriction of sodium intake may remove the fluid, but to lessen the hazard of precipitating hepatic coma it is essential to avoid potassium loss by the concurrent prescribing of potassium chloride solution or other potassium supplements. If hepatic coma occurs, potassium chloride should be administered intravenously. Hepatic coma calls for the immediate elimination of protein from the diet. Laxatives, such as magnesium sulfate, may be used to decrease the absorption of protein.

Sodium acid phosphate (sodium biphosphate) is the drug of choice for patients forming calcium-containing calculi since it depresses serum calcium levels and reduces urinary calcium excretion. Sodium biphosphate is also used as a urinary acidifier in the treatment of an infection or to provide an acid medium for the functioning of a bactericidal or bacteriostatic drug.

If urinary alkalization is called for, sodium bicarbonate is the alkali most commonly employed, but sodium acetate, citrate, or lactate may be used, since they convert to carbonates. The citrate is indicated when protracted alkalization is necessary (as in preventing the formation of uric acid crystals). The lactate, which—like the bicarbonate—can be administered intravenously (the citrate cannot be), is useful in combating the acidosis of diabetes mellitus and uremia. The acetate and the citrate of potassium may be used to increase the alkalinity of the urine, with the advantage that they do not neutralize gastric juice or disturb digestion.

13

Metals in Other Therapies

The skin provides a barrier between the internal organs and the external environment. Its tough outer layer (the epidermis) is a protective covering against the penetration of noxious substances. Skin lesions originate either as a result of injurious contact in the external environment (an infective organism, a toxic chemical, a physical trauma) or a sensitivity response of a particular individual to a particular agent that is harmless to others.

A number of emollients, lotions, and drugs used in the treatment of skin disorders are metal-related.

Aluminum acetate solution (Burow's solution) is employed as a wet dressing in treating a wide range of dermatologic conditions. It is of particular value in treating eczemas. Aluminum subacetate solution, which is basically an astringent and antiseptic wash, is, properly diluted, also used in wet dressings, especially if a somewhat stronger solution is desired.

Cadmium sulfide (in a detergent shampoo base) is recommended for the control of dandruff. Selenium sulfide is even more effective, but some danger attaches to its use because of its high toxicity.

Calcium gluconate is employed to decrease the flow in certain weeping dermatoses. Calcium hydroxide solution (limewater) is included as an astringent in such dermatologic preparations as calamine lotion (which also contains zinc oxide and a small amount of iron oxide). Gold salts are used in the treatment of certain skin diseases, including tuberculous skin disease.

Ammoniated mercury is employed for the external treatment of such skin diseases as impetigo contagiosa, ringworm of the body, fungal infections of the scalp, seborrheic dermatitis, superficial pyodermas, and the scaling of psoriasis. It also combats crab louse infestation. Potassium permanganate solutions provide compresses, wet dressings, and soaks in acute dermatoses, especially when there is secondary infection, but care must be exercised to ensure that the potassium permanganate is completely dissolved; otherwise chemical burning of the skin can result.

Sodium alginate is a basis for emulsions, ointments, creams, lotions, and suspensions used in dermatologic treatment. Titanium dioxide is employed as a protective against sunburn.

As early as the thirteenth century, Myrepsus was employing an ointment containing zinc oxide in the treatment of malignant ulcers. Today zinc oxide is widely used as a dermatologic therapeutic agent, either alone or in a variety of formulations. Its efficacy rests on a combination of protective, mildly astringent, and even mildly antiseptic properties. Back in the 1940s zinc peroxide paste was the therapy of choice in the treatment of indolent (sluggish but not painful) leg ulcers, but there were a number of objections to and restrictions upon its use. In 1944 John R. Cochran of the department of surgery, Northwestern University Medical School, reported on a new type of zinc peroxide ointment that generally overcame the former limitations.[1]

Roughly a quarter of a century later a research team headed by Arthur T. Risbrook of the A. Holly Patterson Home for Nassau County (N.Y.) Aged and Infirm treated "a 79-year-old woman with peripheral arteriosclerosis [and] a large (10 x 17.5 cm) raw painful ulcer of the left leg. She had had the ulcer for five years and, ten months before admission, had undergone an aorto-femoral bypass with sympathectomy. Gold leaf therapy was started but was discontinued after two years for a trial of other methods. These were unsuccessful. After another two-year course of gold leaf therapy, the ulcer was healed. The length of time that elapsed before healing emphasizes the need for perseverence." The authors noted that the therapeutic use of gold leaf dates back to the Digby Receipts (1688).[2]

Zinc stearate is employed alone or in combination. It has the advantages of being a water-repellent protective, of making powder adhere better to the skin, of making pastes smoother and more plastic, and of relieving itching. Zinc stearate is sometimes used as an aid in massage of the skin. Zinc sulfate has been used in various dermatologic medicaments. Combined with sulfurated potash it forms "white lotion" (zinc sulfide) which is employed in the treatment of skin problems.

Eyes

Organic gold compounds have been employed in the treatment of certain types of inflammation of the eye. The antibacterial activity of yellow mercuric oxide (yellow precipitate) ointment has been applied to the treatment of local infections, especially of the eye. White precipitate (ammoniated mercury) ointment is used to treat diseases of the eye. For many years silver nitrate was regarded almost routinely as a prophylactic agent against ophthalmia neonatorum (conjunctivitis in the newborn), its use being mandatory in many states; it is now being supplanted by the sulfonamides and penicillin. Mild silver protein has germicidal action effective against various inflammations of the eye (and of the ear, nose, throat, rectum, urethra, and vagina). Zinc sulfate is employed as an eyewash in the treatment of conjunctivitis and is applied to the skin of the lids in cases of acne, poison ivy, lupus, and impetigo. Zinc oxide is also applied around the lids to combat eczema, impetigo, ringworm, severe itching, and psoriasis. Zinc is chiefly used for its astringent and antiseptic properties.

Antidotes to Poisons

Poisoning can stem from a variety of acts or accidents, including occupational hazards, the bite of a spider, a suicide attempt involving a frank poison or an overdose of medicaments that are harmless as prescribed, and the ingestion of toxic substances (especially in the case of children). Metal compounds are included among the substances employed antidotally, as an expellant or to relieve side effects.

Calcium compounds have been used in the treatment of lead, mercurial, and arsenical poisoning. Where ingested poison cannot be recovered by induced vomiting or gastric lavage, the cathartic of choice (to hasten the passage of the poison through the intestines) is sodium sulfate. Sodium bicarbonate and sodium lactate, by hastening renal excretion of the poison, have aided the treatment of poisoning caused by barbiturates and salicylates. Pain and muscle spasm resulting from the bite of the black widow spider are relieved by calcium gluconate. Sodium chloride and mercurial diuretics promote the excretion of bromide in cases of bromide intoxication. Cyanide combines readily with iron in the ferric (trivalent) state. The most effective method yet devised to treat cyanide poisoning is based on the intentional conversion of hemoglobin iron from the ferrous (with a lower valance than three) to the ferric state by the administration of nitrites. Since speed is essential, amyl nitrite is given by inhalation followed by an intravenous injection of sodium nitrite—followed in turn by an intravenous injection of sodium thiosulfate, which reacts with the cyanide to form an excretable and relatively nontoxic compound. In the case of strychnine poisoning, fatal convulsions may be precipitated by attempts to empty the stomach unless barbiturate sedation is instituted beforehand. Potassium permanganate lavage is then employed. In thallium poisoning, potassium chloride increases the excretion of the metal in the urine.

Burns

Burns have plagued man since he began to use fire and to expose himself to the sun. Modern medicine recognizes four stages in burn therapy: treatment of shock, fluid loss replacement, treatment of the burned tissue and the avoidance of fatal sepsis from the bacteria in the wound, and repair. Prior to the eighteenth and nineteenth centuries, burn therapy was directed toward relief of pain and care of the wound. The idea of repairing the wound with skin grafted from the patient's body did not present itself until the latter half of the nineteenth century, and little attention was paid to the systemic effect of thermal injury until around 1900.

While both shock and fluid loss can prove fatal, especially in third degree burns (and corrective action should be taken immediately), concentration is generally focused on treatment and repair of the skin covering. When the protection of the skin is removed, there is nothing to keep out bacteria that might cause severe infections.

During the first quarter of the present century a variety of therapies for treating third degree burns was introduced. They varied from exposing the burned surface to the air, to the application of medicated waxes and ointments, to soaking in a bathtub for a long period of time. Each new agent or approach was "extolled as the most efficacious from all viewpoints by its discoverer and the myriad of band-wagon-hoppers who almost inevitably follow him." [3] But in the long run no method proved acceptable to all and, whatever the method, infection was frequent and patients, even with only small areas of body surface damage, died.

In 1925 Edward Clark Davidson, a Detroit surgeon, introduced the "tannic acid" treatment, the use of which persisted until 1940. (As early as 1883, a physician named John Dunlop had employed a 6 percent solution of tannic acid on thermal wounds but abandoned its use after a very limited trial.) The treatment involved painting the burned area with a solution of tannic acid. A crust formed when the solution dried. Where less than full thickness of the skin was burned, the coating allowed the skin to heal under the crust; in full thickness burns, the crust separated from the underlying tissue during healing, leaving a prepared surface upon which skin could be grafted. Tannic acid

> was claimed to be superior to any other form of burn therapy: it stopped pain, prevented infection, reduced the loss of fluid and so prevented shock, minimized the need for grafting, allayed toxemia and lessened the danger of cicatricial contracture. We now know that these claims were in general unfounded. However, during the period of enthusiasm for tannic acid, the few informed skeptical voices raised against it were drowned by the symphony of hopefully opinionated sycophants.[4]

Worse than the invalidity of these claims, tannic acid proved to be the cause of many deaths. As early as 1930 John Dunbar of Glasgow, Scotland, noted that the mortality rate of burned patients treated with tannic acid was nearly double that of a control group. His further investigations during 1932 revealed mortality of 60 to 66.7 percent where tannic acid was used as a dressing for burns covering 20 to 40 percent of the body's surface in contrast with 16 percent in like cases where no tannic acid was used.

In the early forties Sumner Koch and Harvey Allen of Northwestern University Medical School declared that burns should be covered with fine mesh gauze saturated with Vaseline, that the gauze should be covered with large amounts of fluffy bandage, and the whole wrapped in roller bandage. This became accepted burn therapy during World War II. Then in 1950 A. B. Wallace, a surgeon at the University of Edinburgh, advocated a return to the earlier practice of complete body exposure to clean dry air. This approach was widely accepted for several years. All this vacillation indicated that there was no ready solution in sight.

About a decade before he introduced the 0.5 percent solution of silver nitrate in the treatment of large human burns, Carl A. Moyer wrote:

> It is nearly 100 years since antisepsis came to dominate the treatment of burns. All that it has accomplished, so far as we can see from data available, has been to offset the good that sound physiological and surgical principles and modern aseptic technic should have afforded. In spite of this, much of our thinking is still captivated by the local antiseptic idea. However, new antiseptic agents are being discarded more rapidly than they were 20 years ago. Within the past decade sulfathiazole, sulfadiazine, Furacin [nitrofurazone], penicillin, and other antibiotic oint-

> ments have been tried but have not received enthusiastic acceptance as agents to be placed on thermal wounds. This may be a sign that the antiseptic idea is about to receive the just interment it has long deserved.[5]

In 1965 he admitted that he had "in the face of deep-seated personal prejudice to the application of antisepsis in treatment of burn wounds" been seeking "a safe antiseptic" for such purpose. He and his co-workers set up criteria, on the basis of which they rejected such materials "previously used as antiseptics on burns" as phenols, mercurials, picric acid, iodoform, antibiotics, soluble sulfonamides, and dyes. Silver salts come close to meeting the criteria.[6]

In 1973 John A. Moncrief of the Medical University of South Carolina offered this evaluation of silver nitrate:

> When the silver nitrate solution is used, the burn wounds are covered with a 5-cm layer of gauze saturated with a warm solution of 0.5 per cent silver nitrate and held in place with light circular dressings. To retain body heat and minimize evaporative water loss, the entire patient is covered with a dry blanket. The silver nitrate is effective not only against pseudomonas but also against the entire spectrum of organisms residing within the burn wound, and clinical bacterial resistance has not been demonstrated. The antibacterial effect resides entirely within the silver ion, which on contact with tissue fluids is precipitated as silver chloride, thereby limiting its effectiveness and precluding any appreciable penetration beyond the surface tissues. Therefore, therapy with silver nitrate has not been as effective in established infections in deeper tissue, and active, vigorous debridement of the burn wound must accompany the use of silver nitrate to permit continued access to the bacterial population. The main hazards of its use are the absorption of large volumes of distilled water from the dressings into the body and the leaching of potassium, sodium, calcium and chloride from the body tissues into the dressing. Replacement by intravenous or oral sodium chloride

and potassium gluconate will readily control these defects when they are used promptly and wisely.[7]

A recent entry into the field of topical burn creams is silver sulfadiazine, which is painless and odorless and produces dressings that do not stick. There is no chemical loss from the body and no staining, as there is with silver nitrate.

In February 1973 a group of researchers from the Experimental Therapeutic Branch of the National Heart and Lung Institute (Bethesda, Maryland) reported on the loss of appetite and distaste for food that reduces the caloric intake of burn patients to a point of insufficiency. With the sense of taste either completely gone or largely distorted, formerly palatable foods taste unpleasant and patients will not eat them.

I. Kelman Cohen and his associates determined that poor appetite and the altered sense of taste were attributable to the heavy loss of bodily zinc that occurs in severe burn cases. Further investigation is being made to determine whether administration of zinc will correct the appetite and taste deviations.[8]

What further research may produce is debatable. Some months after the above report Daniel B. Stone, in his evaluation of oral zinc sulfate quoted earlier, had this to say: "The evidence so far suggests that zinc does not help the burned patient, which surprised me, as a burned patient loses increased amount of zinc in the urine." [9]

Mental Illness

Mental illness demanding therapy on a widespread scale is a product of the twentieth century. It is true that the story of mental disturbance in man is as old as his history and that care (or noncare) of the "feeble-minded" has varied from period to period; but as the world emerged from feudalism into the "Enlightenment," the treatment of "lunatics" (both harmless and dangerous) assumed a pattern that would continue for several centuries. Now the sick wealthy were cared for in their homes, while the sick poor were condemned to hospitals—charitable institutions that too often lacked the cleanliness (and kindness) associated with godliness. The same rule applied to the mentally unbalanced, with the difference that the unfortunate lunatics had also to face cruelty and being made a public spectacle. Despite the efforts of such humanitarians as Rush, Pinel, Tuke, Dorothea Dix, and Freud, the stigma of lunacy was carried into the twentieth century by the mentally ill.[10] Only when, between the wars, the disintegration of the family revealed the number of slightly peculiar members that had been hidden away, cared for, and protected—who could no longer be hidden away, cared for, and protected—did it become apparent that mental illness was not a prerogative of the vicious poor but that practically everyone has been or will be mentally ill to a greater or lesser degree at one time or another. Psychiatrists, psychoanalysts, sanitariums (which replaced insane asylums for those who could afford to pay), with the therapeutic aid of tranquilizers, did their best.

In 1949 an Australian doctor, John F. J. Cade, demonstrated the effectiveness of lithium carbonate in the treatment of mania. In 1969, Ronald R. Fieve, chief of psychiatric research at the New York State Psychiatric Institute, summed up its effect:

> The most unequivocal results of lithium therapy have been in terms of its effectiveness in combating episodes of mania or hypomania. An estimated 6,000 patients have been treated with lithium during the past 20 years, in more than a dozen countries, and by 300 research groups and special lithium clinics. Although not all these patients were presenting symptoms of mania or hypomania at the time of treatment (some were in the interval or

> the depressed phase of manic-depressive illness, while others were presenting symptoms of a wide variety of other disorders), a review of the literature to date indicates that, of all manic patients treated, 20 to 30 percent have failed to respond satisfactorily to lithium, while 70 to 80 percent have shown distinct improvement within one or two weeks. Furthermore, the therapeutic effects of lithium are apparently independent of age, sex, or duration of disorder.[11]

Cade, in his first report on the effectiveness of lithium in subduing acute and recurrent mania, noted that the drug was without effect in depression. Subsequent investigators challenged this conclusion but, says Fieve, lithium's effect "on the moderate to severe endogenous [arising from within the body] depression is at most a weak one; it is clear that this condition is better managed with antidepressants or with ECT [electroconvulsive therapy]. In addition, lithium appears to have no established value in reactive or neurotic depression, or in acute retarded suicidal depression. Whether lithium alone is effective against the mild endogenous depressions that are generally seen by private practitioners, in patients who are still functioning socially and vocationally, also remains to be confirmed." In 1967, a claim was made that maintenance lithium decreased the frequency of hospitalization (and the duration of psychotic episodes) in manic-depressive patients, but, according to Fieve, the claim that the drug is an effective preventive seems premature. As to the effect of lithium on other psychiatric disorders, there have been "isolated reports of successful lithium treatment of the schizophrenias, excited catatonia, paranoid psychosis, confusional excitation, delirious reaction, epilepsy, compulsive neurosis, and other disorders. There have also been claims of lithium's value in treating the premenstrual depression syndrome and adolescent impulsiveness with accompanying mood swings (in adolescents who are disposed to behavior disorder, schizophrenia, or psychopathy). . . . Lithium appears to be of minimum value to none at all against agitations of schizophrenic, neurotic, or organic origin; it may, however, prove to be of value in a limited number of schizo-affective psychoses with a moderate to marked elation component." [12]

One bar to the advance of lithium therapy has been the toxicity of the drug. Toward the end of the 1940s lithium chloride became a popular substitute for individuals on sodium-free diets. Some of these individuals were suffering from heart and kidney disease, conditions in which lithium is particularly dangerous. After four deaths and many serious (if nonfatal) poisonings, the Food and Drug Administration in 1949 banned the use of lithium (except for research purposes), a restriction that was partially lifted in April 1970 to permit its prescription in the treatment of mania. With this in mind, Fieve warns that lithium "is not the 'wonder drug' that some of its more enthusiastic advocates claim it to be. Its low therapeutic index and its toxicity when administered in excessive dosages—presenting the danger of coma and death—are now widely recognized, as are its less serious side effects." Nevertheless, it "appears that, within the next decade, lithium will to a great extent replace ECT at many of the major medical centers, as the preferred treatment for manic-depressive psychosis." [13]

In 1972 Phillip Polatin, chemical director of the New York State Psychiatric Institute, took up the question of clinical versus research applications of lithium carbonate in emotional disorders:

> In clinical medicine, and most particularly in clinical psychiatry, one of the most marked threats of mechanization has come about, ironically, through the great advances of chemotherapy. In this field the results sought frequently obliterate consideration of the identity of the patient as a person. . . . In our studies we have a series of manic-depressive patients treated prophylactically

> with lithium carbonate with excellent results, but, from a clinical point of view, this lithium prophylaxis produced reactions that threatened the livelihood, creativity, or life style of each individual patient so that as a consequence this group elected to discontinue lithium prophylaxis.[14]

Polatin recognizes that the use of lithium "as a preventive of manic-depressive illness, especially the manic type, will undoubtedly become widespread now that it has been released by the F.D.A. for general use by medical practitioners" and freely admits "that Lithium Carbonate prophylaxis has proven of incontestable value in the majority of patients with manic-depressive psychoses, especially in the manic phases, and perhaps in the controversial depressed phase," but he stresses "that it is most important to individualize each patient, to determine whether the prophylactic use of Lithium is really in the patient's best interest, rather than to prescribe Lithium as a prophylactic indiscriminately." [15]

Burton P. Grimes of the St. Peter State Hospital, Minnesota, finds the major benefits of lithium therapy in the control of hyperactivity and the control of side effects. "Manic patients have commented on the freedom from a feeling of fogginess which they had experienced with the large doses of phenothiazines they had required on previous episodes. There is also a longer interval between attacks. . . . The control of hostility is particularly important in the nursing care and supervision of the patient. With lithium therapy, the improved social adjustment in the hospital makes life more pleasant, not only for the patient, but for all the hospital personnel." In summary, Grimes reports that, of fifty-four patients started on lithium for the control of manic and depression reactions, definite improvement was seen in 65.2 percent.[16]

Weighing "the dramatic effects of lithium treatment on mania" against the possibility that lithium may disrupt the sodium-potassium balance in neural tissue and the fact that "lithium is not markedly effective in the treatment of most forms of depression," David Samuel and Zehava Gottesfeld of the Chemistry of the Brain and Behaviour Group at the Weizmann Institute of Science (Israel), raised the question whether "rubidium and possibly caesium might work in the opposite direction and might thereby be more efficacious." By September 1973 they could only report:

> There is little doubt of the effectiveness of lithium therapy on mania but its mode of action is still obscure. . . . The effects, both behavioural and biochemical, of rubidium and caesium are also obscure and further basic work is required before even a tentative mechanism is found for an ever increasing number of empirical observations.[17]

Polatin spoke of patients electing to discontinue lithium prophylaxis because of its threat to their life style, but the concern of Roy G. Fitzgerald, director of inpatient psychiatry at Pennsylvania Hospital, is the many manic patients who, having responded positively to lithium therapy, relapse when they become outpatients because they discontinue taking lithium. He advocates the introduction of "family-oriented psychotherapy [which] can improve the verbal communications between manics and those around them, can help them continue to take lithium, and can help prevent relapses of manic-depressive psychosis, a chronic and potentially fatal illness." Intrinsically, he acknowledges the value of lithium as a medicament, but he does not visualize its acting alone:

> Mania can be seen as a behavioral pattern within manic-depressive psychosis, itself a chronic illness. Both the behavioral pattern and the chronic illness create emotional, social, legal, and biologic disturbance in the individual and those around him (and they may also provoke him!). There are a variety of strategies, all

> of which can play a role in the long-term prophylaxis of this condition. These include vigilance, individual, and family psychotherapy. While lithium is an effective treatment for acute mania and prophylactically for mania and associated depression, its efficacy is enhanced by an understanding of the patient's difficulties, needs, and relations with others. A recent report documents that lithium therapy in the patient can help the spouse and marriage as well.[18]

As early as 1929 a low level of zinc due to excessive urinary loss was detected in schizophrenics. Subsequent investigations have suggested that schizophrenics may have low levels of zinc, manganese, chromium, and molybdenum, high levels of copper, iron, cadmium, mercury and lead. "The last three are, of course, poisons, but the poisoning may produce symptoms that mimic those of schizophrenia. This has been documented for mercury and lead." [19]

In 1967 Carl C. Pfeiffer and Venelin Iliev (also of the New Jersey Neuropsychiatric Institute) found that some schizophrenics had low while others had high histamine levels and turned to trace metal studies to see if abnormalities might be present and might account for these differences. Their extensive exploratory studies of trace metal balance led them to the conclusion that a "probable etiological factor in some of the schizophrenics is a combined deficiency of zinc and manganese with a relative increase in iron or copper or both." [20]

14

Additional Applications of Metals to Modern Medicine

A group of Phoenician sailors beached their boats on the sandy shores of a tidal river and prepared to light a fire. Finding no rocks on which to rest their cooking pots, they used lumps of crude soda ash from the ship's cargo. When the fire died, the ashes contained a shiny substance that became hard when it cooled. So goes the legend of how silicon was first combined with soda ash to produce glass. Whether or not the legend is true, glass was certainly known in Egypt thousands of years before Christ.

The role of glass in medicine has extended from retorts, beakers, pipettes, prisms, and a wide variety of other laboratory gear to syringes, thermometers, ampules, lenses, and containers of all sorts used in hospitals and doctors' offices. In recent years plastics have replaced glass to a marked degree because of their low cost and disposability, but it is inconceivable that glass will totally disappear as a tool of the art of medicine.

Silicon is not the only obscure metal related to glass. Glass filters containing neodymium and praseodymium (mixed together to form didymium) and holmium are used in pharmacologic spectrophotometric tests.

The field is indeed a broad one. Metals (silver, copper, iron, lithium, gold, lead, tungsten, mercury, osmium, ruthenium) are employed in the creation of staining agents to aid diagnostic microscopy; cesium chloride is used in the banding of human chromosomes. But probably the most extensive use of metal in the field of medicine has been in the production of instruments, particularly surgical instruments. In fact, so many instruments are employed today, not only in surgery and obstetrics, for which tools are in a process of constant refinement, but also in general medicine, precise diagnosis, and all aspects of research—and the number is increasing so rapidly—that the mere listing, description, and depiction of them would require several volumes.

Sutures and Clips

Surgical instruments were made entirely of metal until the fourteenth century, when handles of wood and bone appeared; three centuries later these were sometimes replaced by handles of ivory or checkered wood (which offered a firmer grip and prevented slippage). In the last half of the nineteenth century, with the advent of asepsis, the all-metal instrument returned to favor—largely because nobody could be sure that nonmetallic materials would stand up to sterilization by boiling. (At about the same time the introduction of anesthesia expanded the potentials of surgery and brought a demand for more intricate tools—just as, roughly a century later, the demands of heart surgery produced the heart-lung machine.) Today the trend is toward the replacement of metal by plastics, wherever feasible, largely because instruments involving plastics are cheaper to produce, but the demand for instruments made totally or partially of metal is not seriously diminishing. While new varieties of instruments are introduced, existing ones do not become obsolete and their service life is shortened only by neglect, misuse, or the wearing out of moving parts.

Once it has been established that most knives, scissors, forceps, and retractors are made of metal (usually steel) and come in a variety of shapes and sizes, there is little more to be said, but there are certain surgical areas in which necessity has demanded special investigation. Sutures offer a prime example.

Suture materials belong to two main classes—absorbable and nonabsorbable. Catgut, the most common of the absorbable materials, is "hardened" by heating it in a solution of chromium salts when an extended life is desired. Silk is the most popular of the nonabsorbables, but linen thread, nylon, floss silk, horsehair, and stainless steel wire are also used. Both silk and catgut have disadvantages. Silk can be a strong irritant to the surrounding tissues; catgut produces allergic reactions and is subject to fluctuations in its resorption period.

The use of metal sutures in surgery dates from 1858, when the celebrated American surgeon J. Marion Sims (1813–1883) introduced the silver wire suture. Harvey Cushing (1869–1939) extended the use of silver wire clips to brain surgery between 1908 and 1911. Over the years, copper, tin, silver, iron, aluminum, zinc, gold, platinum, cadmium, and nickel were tried and discarded, many of them because they produced an electrolytic reaction which, to a greater or lesser degree, caused interaction with the tissues. Gold and aluminum, while generally inert, were partially dissolved by tissue fluids.

Tantalum, a metal that is elastic, firm, resilient, largely resistent to corrosion, capable of being drawn into threads, inert with respect to living tissues, and free of electrolytic activity, was first used for sutures in Australia in 1922. In World War II tantalum was used in the form of threads for the suturing of nerves and as plates for the closure of skulls. In 1948 scientists in the Soviet Union took up tantalum and devised an apparatus for suturing blood vessels with tantalum clips. Further work by physicians and mechanical engineers extended their use to bronchi, pleura, stomach, intestines, and other tissues. Among their developments was a vascular stapler capable of mechanically suturing the cuffed ends of severed vessels with tantalum staples. (As far back as 1869 Francis H. Brown of Boston drew attention to a *porte-aiguille* [needle holder] for metallic suturing which may be regarded as the grandfather of the Russian stapler.)

In 1972 William M. Moss of the University of California Medical School at Irvine reported that in four hundred bilateral partial vasectomies in which tantalum clips were used, there had been "no infection, hemorrhage, spermatic granulomas, or failures in resulting sterilization." In commenting favorably on Moss's contribution, Rubin H. Flocks, head of the department of urology at the University of Iowa, added:

> The ultimate manner of obstructing the vas deferens [the excretory duct of the testis] may be quite flexible; also it may well be that the kind of clip that is utilized may be readily reopened so that the procedure may be reversible. By applying the clip under vision with the forceps, the amount of external pressure can be varied so that the obstruction can be secured without the complete destruction of the vas deferens.[1]

Another possible method of making vasectomies reversible has been under investigation at the New York Medical Center since 1972—the insertion in the sperm ducts of a tiny stainless steel and gold valve to be turned on or off in a simple procedure requiring only a local anesthetic.

In May 1973 Robert T. Bliss of the University of Cincinnati College of Medicine announced the development of a "true" vasectomy prosthesis that "replaces sutures, devices, clips, clamps, cautery and valves in the performance of a bilateral partial vasectomy. Reliability in maintaining azoospermia is actually increased while improving the possibility of successful reversal." [2] The Bliss Vasectomy Prosthesis consists of two implantable stainless steel cuffs separated by a polyethylene spreader which is flexible enough for easy handling and patient comfort, yet stiff enough to keep the cut ends of the vas deferens apart.

In June 1973 Daniel J. Preston and Charles F. Richards of Wilmington, Delaware, reported on twenty-four years of personal experience (involving 2,000 patients) with the use of annealed stainless steel mesh and annealed stainless steel sutures in the repair of hernias.

> Nearly all hernias can be repaired in some manner without the use of a prosthesis. The recurrence rate can be reduced, however, by using a suitable prosthesis repair in appropriate cases. Hernia tends to develop in a person who has a congenital abdominal wall weakness or who has acquired incompetence of abdominal wall tissues. If these weakened tissues, which have already failed to retain the abdominal contents, are the only supportive structures relied on for repair, then an unacceptable rate of recurrence can be anticipated. Use of steel cloth prostheses for repair is justified by clear evidence of reduction in the recurrence rate which can be attained with it. . . .
>
> Selection of annealed stainless steel fabric as a suitable prosthesis was suggested by prior favorable experience with annealed stainless steel monofilament sutures and ligatures. This suture material was used long before mesh implant repair was considered. Annealed stainless steel has proven to be inert in living tissue, has permanent strength and durability, does not have a tendency to work-harden, and is malleable. . . .
>
> In the more than 2000 repairs that we have done, the incidence of infection is 0.1 percent. In reviewing all our cases, we found

none that had become infected primarily. Stainless steel mesh has been put in wounds that have been infected at the time of its insertion without the formation of sinus tracts or abscesses. The infected cases have been evisceration of abdominal wounds, or recurrent abdominal wall hernias following the use of tantalum where small bowel fistulas had developed. In no instance has it been necessary to remove the steel cloth implant because of wound drainage or infection. . . .

Wire mesh implants and stainless steel sutures are not foolproof and do not guarantee perfect results, but when they are used with reasonable skill and judgment, it is believed that better results can be obtained with them in appropriate cases.[3]

Needles

"Hollow *needles* are made of steel. Their purpose is to carry a fluid substance from one region to another. Attached to a syringe they can be used for injecting fluid into the body or extracting fluid from it."[4] However, one form of needle, the triangular tipped trocar (with its metal cannula or sheath), introduced in 1851 by Charles-Gabriel Pravaz (1791–1853) of Lyon, France, for the injection of iron perchloride into aneurysms to induce clotting, was made of gold or platinum, with a syringe fashioned of silver. The hypodermic needle and syringe (with cannula) to relieve painful afflictions of the nerves was the 1845 invention of Francis Rynd (1801–1861) of Dublin, but his report had gone largely unnoticed.

In 1900 J. Leonard Corning (1855–1923) of New York City described an instrument he had designed for spinal anesthesia. It was "3½ to 4 inches long and was made of gold or of 'platina' thus rendering it bendable but unlikely to break."[5] An additional reason for preferring the relatively expensive needles of gold, platinum, and platino-iridium was the predisposition to rusting of low-carbon steel. Such rusting might escape detection until an actual break occurred. Even "steel nickeled over" blackened from constant boiling and lost its polish, exposing the instrument to corrosion which sometimes choked the needle. In 1926 W. R. Meeker of the Mayo Clinic expressed a preference for lightweight spinal puncture needles made of nickel alloy, that would bend but not break, over platinum, platino-iridium, and gold needles which, in addition to being costly, were difficult to keep sharpened. Gaston Labat (1877–1934), a specialist in nerve blocking and author of the pioneer work, *Regional Anesthesia* (1923), who received his M.D. in Paris in 1920 and two years later moved to the United States, favored a spinal puncture needle of nickel with a stylet to increase rigidity. George P. Pitkin of Holy Name Hospital (Teaneck, New Jersey) used needles of nickeloid or rustless steel but favored the latter because he found that the rustless steel needle was less likely to break, retained a sharp point longer, and withstood considerable manipulation without bending, while some nickel-plated needles tended to "peel" and did not hold a point.

In the early 1920s a heat-tempered stainless steel that was rust-resistant but not rustproof was developed in England. About the same time, the German steel V2A, produced by Krupp, became available and was adopted for needle manufacturing. "Thus the low-carbon steels were displaced by the higher-carbon, rust-resistant products; and these, in turn, were replaced by the safer, rustproof, corrosion-resistant needles." The initial cost of needles made from corrosion-resistant metals was high, but their cleanliness added a safety factor. By the mid-1940s most American needle manufacturers were using improved V2A steel.[6]

In 1939 William T. Lemmon of Jefferson Medical College (Philadelphia), in introducing "continuous spinal anesthesia," employed a German silver (nickel-silver alloy) needle, but refinements in the science of metallurgy later allowed his needle to be produced in hyperchrome stainless steel specially annealed to render it malleable. The past decade has seen the introduction of disposable needles manufactured of stainless steel in conformity with federal specifications.

A different form of needle—a solid needle—is employed in acupuncture, a practice that appears to have been introduced by the Chinese emperor Huang Ti (2697–2597 B.C.). Its objective is to bring the male yang and the female yin principles into the equilibrium regarded as necessary to health. It involves pricking the body with needles, coarse or fine, long or short, at various points, sometimes twisting the needle in the stretched skin. The original needles were of gold and silver, possibly copper. Later iron, steel, and brass were employed. Around A.D. 1683 a Dutch surgeon named Ten-Rhyne introduced acupuncture into Europe, where it was regarded as a panacea and was widely practiced, especially in France.

Today acupuncture is enjoying a rebirth as an alternative to drug anesthesia in the relief and avoidance of pain.

> Since the reported success of the use of acupuncture analgesia in several hundred thousand operations in China, great interest in this type of analgesia has been aroused all over the world. In Japan, America and other nations, medical people have begun to investigate the application of acupuncture analgesia in their daily medical and clinical work. . . .
>
> Acupuncture analgesia . . . is based on Chinese traditional medical theory, *Ching-lo,* which states that by applying pressure to certain specific points on the body, these points (on the meridians) will become numb. Pressure is applied by using a certain method of needling so that pain is either dulled or removed altogether. Thus, in using this technique during surgical operations, sensation in the area in which the operation is to be performed can be dulled while the patient remains entirely conscious.
>
> At present in China this technique is widely used in both urban and rural hospitals and in mobile medical clinics. It has been used for surgery on the brain, neck, chest, abdomen and

> limbs. In addition, it has been applied in obstetrical operations and in operations on the ears, nose and throat. Acupuncture has attained a definite place in Chinese medical practice and, when it can be used appropriately, has become the first choice of anaesthetic.

But there are limitations. Acupuncture analgesia is still in the early stages of its modern development. Since it does not dull the nerves as completely and as effectively as drugs, it is not suitable for every operation. But if it is regarded, as it should be, as a new technique in anesthesia and is employed in conjunction with drugs, it can be viewed as a useful addition to the anesthesiologist's armamentarium.[7]

Today's acupuncture needles are made of stainless steel, silver, or gold.

Crushing and Exploring Instruments

Stone in the bladder has been known since the earliest times. A stone in the pelvis of a skeleton in a predynastic Egyptian grave at El Amara has been estimated to be more than 7,000 years old. Instruments for removing or crushing stones have had a long historic development and certainly involved metal, but our interest at this point is centered on diagnostic aspects—the detecting and pinpointing of stone. Prior to the nineteenth century reliance was placed on the bladder sound (or probe), sometimes equipped with a sounding board which enabled the investigator to listen to what was going on in the organ. But early in that century the need was recognized to see what was going on rather than just feel or hear it.

An endoscope is any instrument for viewing the interior of a body cavity through its natural opening—for example, the bronchoscope, the gastroscope, the sigmoidoscope, the anoscope, the proctoscope, and the cystoscope. The cystoscope is the instrument employed to detect stones and other diseases in the bladder, with entry by way of the urethra. One of the basic problems in developing such an instrument was providing an adequate light source and directing it to the area under observation.

The earliest attempt at cystoscopy was made by Philipp Bozzini (1773–1809) of Frankfurt am Main whose objective was to extend the principle of reflected light, long employed in the ear speculum, to the urethra. His *Lichtleiter,* or light conductor, reported in 1807, was made of silver protected with sharkskin. The source of light was a beeswax candle held in position in one half of a boxlike chamber, the light from which was reflected down the examining tube by a silver mirror. "The observer's eye, focused at the back of the box, could see by the reflected light through metal tubes inserted into the urethra." [8] Bozzini's tube was not very practical and the same has to be said of numerous creations, exhibited in the next half-century, that relied on reflected light, including those of Pierre Ségalas (1792–1875) of France and John D. Fisher (1798–1850) of Boston.

In 1865 Antonin Desormeaux (1815–1882) of Paris published a book on endoscopy and designed a cystoscope with a lamp that burned terebinth (a form of turpentine), alcohol, and paraffin, but "the light source was never adequate for visualization of the vesical wall even at maximum intensity." Still his book "helped to popularize the use of the cystoscope, and it is in this role of popularizer rather than in the role of innovator that Desormeaux might deserve the title of 'Father of Endoscopy.' " [9]

In 1867 Edmund Andrews (1824–1904) of Chicago attempted a step forward by introducing magnesium wire into the flame of Desormeaux's oil lamp, finding, in his own words, that magnesium wire "burns with a white light, whose brilliancy dazzles like the glare of the sun at noonday." With this light he felt that the interior of the urethra "could not have been seen any better, had it been dissected and laid in the sunlight." [10] Nonetheless his magnesium wire failed to give sufficient light. He tried a long row of gas jets but, when these proved inadequate, "he devised an auscultation sound which greatly increased sensitivity in detecting tiny stone fragments in the bladder." [11] With the failure of Andrews's effort, attempts to achieve endoscopic lighting using an external source dwindled.

It was a dentist in Breslau, Poland, who conceived the idea of an intratubular (internal) light source. Julius Bruck (1840–1902) was thoroughly appreciative of the possible applications of endoscopy. In 1867 he introduced an instrument lit by an electrically heated platinum loop. Inserted into the rectum, its light transilluminated the bladder; by simultaneously passing a tube into the bladder, Bruck was able to see the mucosa of the posterior wall. To compensate for the heat generated and prevent the patient from being burned, the filament had to be encased in a water bath. Despite this problem, and early inefficiency, the road to the internal light source had been opened. Then in 1877 "the potential of Bruck's theoretical advance was realized in other advances which virtually established the form of a clinically useful cystoscope as it is used today. The remarkable man responsible for this synthesis was Max Nitze." [12]

Max Nitze (1848–1906), of Dresden and subsequently Vienna, recognized two basic problems—insufficient illumination of the area under inspection and minuteness of the field of vision. His solutions lay in putting the light source into the cavity and employing a lens system which could be more sophisticated because the light for viewing no longer had to traverse the optical system from outside to inside. Among the benefits were greater magnification and greater angle of vision. Nitze employed Bruck's platinum loop heated by a direct current and cooled by constantly circulating water.

The next thirty years were largely devoted to attempts (by Nitze and others) to find a reliable light source. Fusion of platinum loops was apt to terminate an examination prematurely. Edison's incandescent lamp (1880), which first employed a metallic filament and then a carbon one, lessened the problem, but burned out lights have continued to plague cystologists.

A real change came in 1909 when William Otis and Reinhold Wappler introduced an instrument in which the European prism was replaced by a hemispherical lens that increased the viewing area in the bladder about four times. Within the past twenty years a new direction has been taken with the introduction of fiber optics as the means of transmitting light to the interior of the bladder. On 26 October 1972 David M. Wallace, president of the urology section of the Royal Society of Medicine, concluded his address to the section with these words:

> Strange though it may seem, the principle of fibre optics was first announced to the Royal Society in London where Tindall demonstrated in 1872 that, by using internal reflection, light could be made to bend round corners. However, little was done about this until 1951 when Hopkins introduced the principle of coating fibres of glass with an outer layer of glass which had a different refractive index. This ensured internal reflection of the light. The introduction of these glass fibres as noncoherent fibres has done more to facilitate the examination of the bladder than many inventions of this century. By bringing the light back to the operator's end of the cystoscope we have completed a full circle so that we are once again in the era of Bozzini where the light is placed outside the patient and is transmitted into the bladder by physical means. The most recent evolution of the optical system was again due to Hopkins who in 1956 demonstrated to the British Association of Urological Surgeons in Glasgow his new solid rod lens system, which has subsequently been adopted by Stortz. This lens system, which is yet another milestone on the road to perfect endoscopy, used solid rods of glass with small spaces be-

tween them to act as a lens. The improved optics of this system has contributed markedly to better visualization and even to photography of the bladder.[13]

Vacuum Extractors and IUDs

The idea of "suckers or tractors applied to the head with the object of removing depressions of the cranium in children," [14] dates back to Ambroise Paré (1510–1590): "If the bones do not spring back of themselves, you must apply a cupping glass with a great flame, with all command to the patient to force his breath up as powerfully as he can, keeping his mouth and nose shut." Then in 1632 Fabricius Hildanus (1560–1634), the father of German surgery, wrote: "Depressed fracture of the skull in infants with soft bones may be brought up with a leather sucker." In 1706 James Yonge (1647–1721), mayor of Plymouth, England, and surgeon to the city's naval hospital, reported a case in which "a cupping glass fixt to the scalp with an air pump failed to draw out the head." Almost ninety years later, J. F. Saemann of Jena "saw in a dream an air pump wherewith one can seize the head of an infant without injury to mother or child. The pump was made of brass and had a covering of rubber." In 1829 Scotch-born Neil Arnott (1788–1874) published in London his *Elements of Physics or Natural Philosophy* in which he advocated the use of a pneumatic tractor which seemed "peculiarly adapted to a purpose of obstetric surgery, viz., as a substitute for steel forceps in the hands of men who are deficient in manual dexterity, whether from inexperience or natural ineptitude" and might "assist in raising depressed portions of fractured skull." [15] But "the first instrument developed and made practicable and useful was presented by *James Young Simpson* (1849) in Edinburgh." [16]

The idea of an air tractor occurred to Simpson as early as 1836 when he saw a group of boys "lifting large stones with round pieces of leather wetted commonly called suckers." [17] In suggesting the utilization of this phenomenon in medical practice, Simpson was aware of the earlier pronouncements of Paré, Hildanus, and Arnott.

Simpson's first cup, produced in 1849, consisted of a trumpet-shaped concave disc, covered with leather at its broader end and fitted with a piston. By 1855 he had found the most effective obstetrical tractor to be "a slender short brass syringe, 1½ or 2 inches long, worked by a double-valved piston, like a breast-pump, having attached to its lower extremity a cup of half an inch in dept, [*sic*] and 1½ inches broad at its edge. Over this inner cup was placed a second cup formed of vulcanized caoutchouc [rubber], and so deep as to overlap the edge of the inner by six or eight lines [about one-half inch]. The mouth of the inner cup was covered by a diaphragm of very open brass wire gauze, and over it a piece of thin sponge, flannel, or the like, was placed, with the view of preventing injury to the scalp, and not allowing it to be elongated and drawn up into the vacuous space." Simpson considered his tractor superior to forceps which "in inexperienced hands" could prove injurious to the mother or the child.[18]

The next hundred years saw a dozen variations on the Simpson designs. Then in the 1950s Tage Malmström of Gothenburg, Sweden, produced "the most successful instrument designed up to the present, . . . which is now so extensively used all over the world." [19] In his original vacuum extractor (1953), a traction chain attached to a metal handle passed through the rubber suction tube to two metal cups (the outer one perforated) backed by rubber. The metal handle had a connector for a tube to the suction pump. In a 1956 model, Malmström reduced the diameter and curve of the metal cups "because of the properties of the scalp membranes." In his latest (1967) model, "the metal pipe to which the traction tube is fastened has been placed in a depression of the cup. Thus the cup has become shallower, making its introduction into the vagina easier. In addition, the metallic bottom plate (VE/56) has been replaced by a silicon rubber plate. The cross-formed traction handle of the model VE/56 has been replaced by a simple small pipe with a metal hook. In the VE/67 the steady traction handle can be placed on the traction tube at the desired distance from the foetal head." [20]

The use of trace elements to regulate fertility is not a new approach. As early as 1853 the French anthropologist-physician Armand de Quatrefages (1810–1892) showed that copper salts were toxic to sperm, and investigations a century later suggested that the effect might result from the introduction of magnesium chloride, sodium silicate, chromium trioxide, cadmium chloride, or lead nitrate. In 1969 Jaime A. Zipper and his associates at Catedra "E" de Obstetricia, Hospital Barros Luco-Trudeau and Instituto de Fisiologia, Universidad de Chile, Santiago, tested copper, zinc, magnesium, tin, and silver and arrived at the conclusion that copper and zinc were the most effective.

The use of plastic intrauterine contraceptive devices (IUDs) has become widespread over the past twenty years, sufficient time and experience in which to determine that IUDs made solely of inert material are of questionable effectiveness. Their

> use has undergone successive waves of enthusiastic advocacy followed by periods of discouragement and disparagement when clinical experience has failed to bring out initial high expectations. . . . Because such devices could be produced simply and inexpensively, could be inserted by paramedical personnel trained in a relatively short time, and because they had the potential for remaining in place for months and years due to their inertness to tissue, there were high hopes that this method could be the near ideal contraceptive for use in bringing down fertility rates sharply in the lesser developed countries of the world. Unfortunately, extensive experience in Taiwan, Singapore, India, Thailand, etc.,

> failed to bear out these expectations. Even though many different shapes and sizes were devised and used, the high frequency of side effects, uterine cramps, menorrhagia and expulsion, resulted in a disappointingly low retention and continuation rates [sic], and the IUDs lost popularity.[21]

Nevertheless the IUD remained a desirable contraceptive method and recent investigation has been in the direction of adding to the plastic an "active" element to contribute to the contraceptive efficiency of the device.

The idea of using metal to prevent conception dates back to the 1930s when gold was employed in IUDs. But the venture was not pursued and it was over thirty years before Zipper and Howard J. Tatum produced their "copper T" and then their "copper 7" device, the latter seemingly having become the device of choice. Like the "copper T" before it, the "copper 7" derives its name from the shape of the device. It is made from a copolymer of polypropyrene and polyethylene which has copper wire (0.2 mm in diameter and 32 mm in length) with a surface area of approximately 200 mm wound round the stem. Barium sulfate is incorporated into the plastic to make it radio-opaque.

In 1972 a Canadian test of the T-Cu involving 518 patients led to this conclusion:

> Expulsions, removals, and numbers of pregnancies are all less than for other IUDs except the Dalkon shield. The experience reported for the Dalkon shield is of short duration and could yield apparently better results that are misleading.[22]

Clinical experience with the Cu7 has been reported from Britain (342 women—1196.5 women-months) and the United States (186 women—1693 women-months).[23] The major advantages of the Cu7 device over the classical IUDs seems to lie in a reduction of accidental pregnancies to about half the number previously occurring, ease of insertion, a very low incidence of menstrual disturbance in the first three months after insertion, a very low rate of removal for medical reasons, and fewer expulsions.

Furthermore, according to John Swanson and his colleagues at the Mount Sinai School of Medicine, New York, the expanding use of copper-containing IUDs could help to stop the alarming spread of gonorrhea. Unlike most bacteria, gonococci are particularly susceptible to minute quantities of copper. Swanson has suggested that the small amounts of copper salts solubilized from IUDs are a sufficient prophylactic.[24].

Instruments for the Ear and the Eye

Otosclerosis is a chronic progressive disease involving hearing loss as a result of impairment of normal vibration and transmission of sound. Surgical treatment for progressive otosclerotic deafness "has been described in the medical literature extensively for over one hundred years. Perhaps the ancients, too, exorcized their devils of deafness by similar means." [25] However, real progress was not made until 1930s, when the fenestration principle (cutting a window in the labyrinth of the ear), which had been introduced in the nineteenth century and then abandoned, was readopted. One approach involved closing the opened semicircular canal with gold leaf kept in place by a graft of adipose tissue. In 1952 it was recognized that, while fenestration was the essential procedure when the stapes (or stirrup) in the ear was totally fixed, a surgical attempt to mobilize the stapes should be made when it was only partially fixed. If successful, this operation could obviate the need for the more complex fenestration procedure. Several methods of mobilization were devised, one involving the insertion of a tantalum pin and another—used only in extreme cases—involving the total removal of the stapes and its replacement by a plastic prosthesis held in place by tantalum wire.

In 1971 John J. Shea of the University of Tennessee at Memphis presented long-term results for several surgical procedures and concluded that the "large fenestration-teflon piston with skin graft" had been his best technique to date, but two years later Francis A. Sooy and his associates, of the University of California School of Medicine at San Francisco, reported an eight-year study of the use of a preformed stainless steel wire prosthesis in stapedectomies, stating that their "findings to date strongly supported the use of wire-vein graft technique in stapedectomy for otosclerosis." [26]

Hugh Beckman and H. Saul Sugar of the department of ophthalmology and research, Sinai Hospital, Detroit, have been doing some interesting work on the eye involving neodymium—one of the rare earth metals. In 1972 they reported on the use of ruby laser energy in treating glaucomatous eyes. Subsequently they "speculated that neodymium energy, which is less absorbed by uveal pigment, might better penetrate into the inner layers of the ciliary body, and thus possibly be more effective than ruby energy in more deeply pigmented eyes." They modified their ruby laser system by replacing the ruby rod with a ⅜-inch neodymium rod. Eighteen glaucomatous eyes were treated with the neodymium laser energy of which thirteen were considered to be treated successfully; this compared favorably with twelve successes out of twenty when the ruby laser was used.[27]

Treatment of the Skin, Fistulas, and the Mouth

In 1973 Leon Goldman and his colleagues at the laser laboratory of the Medical Center of the University of Cincinnati announced that a high-power Neodymium-Yttrium Aluminum Garnet (Nd-YAG) laser with a special fiber optics scalpel had been used "for the first time for surgery in man." Primary experimentation involved patients with skin problems, five with tattoos, three with port-wine marks, one with blue-rubber-bleb angioma, and three with multiple malignancies of the skin. Preliminary results indicated that this laser can be used with complete ease, flexibility, and precision for specific areas. Moreover, the high-power output Neodymium-YAG laser provided superior arrest of hemorrhage in tissue. "The Neodymium-YAG laser is now being developed for incorporation into an operative laser endoscope for use in cavities of the body, for metallic dentures in the mouth, and also for laser transillumination studies for detection of masses in different densities in the skin and soft tissues." [28]

A variety of techniques has been used, with varying success, to close fistulas occurring between the mouth and the maxillary sinus. In cases where the fistula remains open due to persistent infection in the maxillary sinus, proper cleansing of the sinus with an antibiotic irrigation solution introduced through the fistula and creation of additional drainage via a nasoantral window may be all that is needed to ensure closure. Beyond this, various surgical techniques that have been employed have proved reliable with small oroantral fistulas but unpredictable in very large fistulas. It is for these difficult cases that William Meyerhoff and his associates at the University of Minnesota Medical School have developed their gold foil technique.

This surgical procedure involves denuding the margins of the oroantral fistula of all mucoperiosteum and covering the opening with an overlapping margin of 36-gauge gold foil. The mucoperiosteal flaps elevated to expose the bony margins of the fistula are then sutured loosely over the foil to act as an envelope holding the gold in place. Mucosa from the maxillary sinus migrates across the inner surface of the foil, providing closure on one side. Gradually the mucoperiosteal flap retracts, exposing the gold foil but still maintaining it in its proper position. After several weeks the foil is removed showing an intact sinus mucosal lining.

> We have had successful results in seven patients treated by this method, some of whom were failures of a standard flap technique and all of [whose fistulas] were quite large. The advantages of this procedure are that it is a safe, simple and physiologic method of closing oroantral fistulas without distorting intra-oral anatomy, and we feel that it should receive wider attention by otolaryngologists in the future.[29]

The use of gold leaf as an adjunct to closure of vesicovaginal and vesicorectal fistulas has been reported by Fletcher C. Derrick, Jr., of the George Washington University Medical Center, Washington, D.C.[30]

Maynard Ferguson, a well-known horn player, was long plagued by damage to the oral tissues resulting from constant contact with metal surfaces. His long-time friend and physician, James T. McClowry of Springdale, Pennsylvania, decided on the use of vitallium (an alloy of cobalt, chromium, molybdenum, manganese, silicon, and carbon) because of its low reactivity to human tissue. "An extremely hard metal, it apparently qualifies chemically as well as from a solubility and permeability standpoint as one of the 'safer' alloys to contact oral tissue and saliva, as it is widely used in dental prosthesis." However, to make mouthpieces entirely of vitallium would be a costly procedure—and an unnecessary one because of the wearability of the alloy. A coating of existing gold, silver, and alloy mouthpieces was all that was necessary—a coating that could be repeated in the unlikely event of wear. To solve the problem of how to apply the coating, McClowry turned to an "ion spluttering" technique developed by NASA for the coating of metal surfaces in space vehicles.[31]

15

Bones, Joints, Implants, and Prostheses

"In dealing with the vast and complex problems of reconstructive bone surgery and some of the more complicated fracture problems, one of the surgeon's greatest allies is the use of metal for internal fixation or replacement of portions of bone. This requires that surgeons understand and be able to apply both the biological and engineering principles involved. The biological and engineering principles, however, frequently come into conflict, and under these circumstances a compromise must be made to give the best possible results under the circumstances." [1]

The use of iron wire to fix bone dates from the 1770s, but for nearly a century (until Lister introduced and advocated antiseptic precautions in surgery) the use of metal implants in fracture cases followed a stormy path, the metal being too often blamed for the fatal infections that stemmed from surgery. As a result, the surgeons failed to differentiate between tissue reaction to and tissue tolerance of various metals. In 1829 a study by J. Levert of tissue tolerance to suture wires made of gold, silver, lead, and platinum revealed that platinum was the least irritating, but in 1844 Joseph Pancoast (1805–1882) of New Jersey declared in his *Treatise on Operative Surgery,* published in Philadelphia, that metal implants should not be used in fracture cases because they caused bone destruction. In 1847 Joseph-François Malgaigne (1806–1865), described by John Shaw Billings (1838–1913) as "the greatest surgical historian and critic the world has yet seen," proposed a compromise between those who advocated open and those who advocated closed reduction. "He introduced an external-internal fixation unit for fractures of the knee and elbow, in which adjustable hooks were attached to a clamp. The hooks went through the skin into the bone fragments and the external clamp was tightened using a special wrench. Although this particular device had only short-lived success, orthopedic surgeons currently use modifications of this procedure." [2]

The insertion of metal plates was introduced in 1886 by W. Hansmann. The screws for the plate and one end of the plate itself protruded from the wound so that the plate could be easily removed after four to eight weeks, using a watch key. Over the next fifteen to twenty years plates were made of a variety of metals. At the turn of the century Albin Lambotte of Brussels advised brass plates. Between then and 1909 he experimented with aluminum, silver, and red copper (all of which proved too malleable) and with magnesium plates with steel screws. Magnesium disintegrated so rapidly that the metal had disappeared before the fracture was repaired. He finally settled on soft steel plated with gold or nickel.

The Scottish surgeon William Arbuthnot Lane (1856–1943) insisted on rigid antisepsis (the Lane "no touch" technique). It was his claim that he had never observed rarefying osteitis (increased porosity of bone) in any of his patients, which led him to the conclusion that "rarefying osteitis in plain English means dirty surgery and is a useful term to cover surgical incompetence." [3] He produced rigid plates of tool or crucible steel that could be left inside the wound. The Lane plate was later condemned by W. D. Sherman of Pittsburgh as being too brittle to withstand bending and twisting strains. After three years of research Sherman, in 1912, developed a new steel alloy that contained vanadium and chromium in addition to carbon. It met his definition of ideal steel for a bone plate:

> one that has a sufficient elastic limit with greatest ductibility so that in case a strain should be exerted we would have a bending of the plate instead of a break. . . . The addition of vanadium to a high carbon steel intensifies the hardening elements making the steel more dense and tough thereby increasing the elastic ratio: *i.e.,* ratio between elastic limit and elongation. . . . It would take great force to bend a vanadium plate sufficiently to break it.[4]

In the course of his investigation he rejected silver as having too low an elastic limit and aluminum because it lacked bending properties. About the same time Ernest William Hey-Groves (1872–1944) of Bristol, England, reported that nickel-plated steel had no irritating effect on the tissue (a view that was challenged sixty years later when nickel dermatitis was observed in a number of prosthesis cases) and that magnesium, which was rapidly absorbed, was a powerful stimulant to bone formation.

"Having had much experience in the various methods suggested for the fixation of bones after operations and injuries," wrote Thomas Annandale of Edinburgh University in 1897, "I have come to the conclusion that one of the most simple and efficient means of obtaining such fixation is the employment of steel pins. Should the portion of bone involved in this procedure be in a sound condition the success of the fixation is almost invariably secured." [5] But Annandale's success and recognition evidently was short-lived, because in 1908 Hey-Groves was credited with introducing intramedullary nailing for fractures of the shaft of long bones and of the neck of the femur. Even then, many problems developed with the materials employed and the procedure went into abeyance until the 1930s and the availability of stainless steel.

> The modern era of metallic implants has been characterized by the development of more acceptable alloys and by advances in the mechanical engineering of prosthetic appliances. In 1936 a new alloy of cobalt, chromium and molybdenum (vitallium) was shown to be essentially inert in body fluids. To date, over 15 million vitallium appliances have been surgically implanted. Newer minimally-reactive stainless steels have also been developed. With the development of image intensification in the 1950s, reduction and internal fixation of fractures could be observed readily without subjecting patients to high doses of radiation. Modern implants have now been improved to the point where they are used with almost complete confidence by orthopedic surgeons. [6]

Vitallium plates (12, 18, and 25 millimeters long and 4 millimeters wide) are being successfully used in maxillofacial surgery.

Orthopedic Replacements

> In orthopedic circles, the Year of the Hip has been replaced by the Year of the Knee. Building on the knowledge gained from total hip replacement, surgeons are setting their sights one notch lower, and are learning to replace the entire knee joint with metal and plastic devices. [7]

Arthritic hip disease has long challenged the orthopedic surgeon. At the beginning of this century arthroplasties of the hip were attempted. The procedure involved smoothing the head of the femur and the acetabulum (the cup into which the head of the femur fits), using fascia as an interposing membrane. Toward the end of the 1930s stainless steel replaced the fascia. A decade later a stainless steel "endoprosthesis," originally devised to treat a fresh fracture of the neck of the femur, was adapted to the treatment of painful hip disease. Unfortunately all these procedures called for prolonged postoperative physical therapy with an uncertain outcome. Other early procedures relieved pain but at the cost of hip joint instability and a shortened leg or a total loss of motion. In recent years, however, the development of refined and machined plastics and metals has led to total hip replacement by an appliance involving a prosthetic femoral head and a prosthetic acetabulum. Of the three possible combinations—a metal femoral prosthesis articulating with a metal acetabular component, a metal acetabular prosthesis articulating with a plastic femoral head, and finally a metal femoral prosthesis articulating with a plastic acetabular component—the last has won almost universal acceptance. The femoral component is made of highly refined steel with a highly polished head that is cemented into the medullary canal of the femur. (A metal wire in the rim of the plastic cup allows X-ray visualization and the cement that holds the cup to the pelvis contains barium sulfate for the same purpose.) To date there has been no adverse reaction of the tissue to the foreign materials. "In fact, the acceptance of these materials by the living tissue has been quite amazing, and the likelihood of an untoward reaction by the body appears such a remote possibility that it is not a deterrent in selection of the patient for hip replacement. . . . Although several years will be needed to thoroughly evaluate the results, at present this operation is considered one of the most useful procedures in all of orthopedic surgery." [8]

The few low-grade infections that have developed have been attributed to bacteria in the operating room air. While such infections occur in only about 1 percent of the cases, the

Hollywood Presbyterian Hospital in Los Angeles in 1970 took to "tenting" the area surrounding the operating table, introducing a change of ultra-filtered air every six seconds and garbing the doctors and operating room staff in "space suits." The extent to which such extreme caution is necessary is open to question. Cyril P. Monty and A. K. Sahukar, reporting to the Royal Society of Medicine in October 1972 on fifty hip arthroplasties performed over an eleven-month period, reached the conclusion that "no antibiotics are necessary either pre- or post-operatively, although a topical wash during the procedure is most helpful as both a mechanical cleanser and antibacterial agent" and that metal to plastic prostheses can be carried out "as routine cases on the operating list of a busy orthopædic department using the theatre facilities of a district general hopital." [9]

> Knee prostheses fall into two main categories: hinge joints, which, as the name implies, can move in only one plane; and a variety of devices which attempt to simulate the anatomical functions of the normal knee. In these, the condylar surfaces of the distal femur are replaced—usually by metal—and the proximal surface of the tibia is cut back to accommodate what might be called "tracks" in which the prosthetic condyles move. The tibial portion of the prosthesis is most commonly made of high-density polyethylene. Both parts are cemented with methylmethacrylate.
>
> The choice between using a hinge or one of the other devices depends on the degree of damage already present in the joint.[10]

In 1973 artificial knees were still regarded as far from perfect. Unless the patient's ligaments were already destroyed by disease, the "end-to-end" prosthesis (which, not being intrinsically stable, required competent ligamentous structures) has generally been regarded as preferable because the hinge knee cannot flex with side-to-side stress and because, in the event of failure, the surgeon can rarely, because of the bone removed, permanently fix the knee as a salvage procedure, as can be done when the end-to-end prosthesis is employed. (While the rigid knee is less than satisfactory it is preferable to amputation.)

An early starter in the race to produce total knee prostheses was the knee hinge developed about twenty years ago by Börje Walldius of Stockholm. Designed for use without bone cement, it was originally made of acrylic, but when this proved too weak, Walldius substituted stainless steel, switching in 1958 to vitallium, after which he was able to report success in 80 percent of his cases. In introducing a total knee replacement symposium in the July–August 1973 issue of *Clinical Orthopaedics and Related Research,* guest editors Donald B. Kettlekamp and Robert B. Leach had this to say: "The following articles on total knee replacement arthroplasty range from the considerable experience already gained with the Walldius prosthesis to the Spherocentric knee which is only now ready for clinical trial. In addition to the prostheses presented here, there is at least one additional type commercially available in this country and myriad others in various stages of design. Prosthetic selection for a given patient is complicated by lack of hard clinical data secondary to small clinical series, and short follow-up for all but the Walldius prosthesis." [11] This is important because the "mechanical hinge, though an accepted surgical method, has fundamental limitations . . . which cannot be overcome." Consequently the "development of newer designs and principles should be interesting to evaluate in the near future." [12]

Late in 1973 surgeons and biomedical engineers of the University of Cincinnati Biomechanics and Biomaterials Laboratory reported that they had successfully implanted an artificial thumb joint in a 54-year-old man. "Half ounce in weight, it is the size of an average thumb joint and operates like a simple hinge with a cotter pin to lock it. It is made of vitallium, a cobalt alloy. A British-developed cement was used to fix the joint in place." [13]

> The use of metal alloys for nails, plates, and formed substitutes for bones and joints is decidedly widespread. Virtually all types of the common metals and alloys have been tried in prosthetic devices; gold, silver, copper, lead, zinc, cadmium, tin, iron, nickel, aluminum, magnesium, vanadium, bronze, brass, steel alloyed with other metals, ticonium. Vitallium, titanium, zirconium, and various types of stainless steels have been employed at various times and in various shapes in hopes of improving bone healing. The most common requirements are corrosion resistance and general inertness.
>
> The three most common metal systems currently employed are the cobalt-chromium base alloys, some of the stainless steels, and unalloyed titanium. None of these materials duplicate the low-friction-bearing surfaces at the joints, and they often fail to remain firmly attached to the shaft of the bone. In addition, the electrolytes in body fluids can attack the surfaces of the prostheses made of stainless steel or vitallium. . . .
>
> New techniques for developing porous metal may have applications for the production of bone prosthesis. For example, powdered metals cam be tamped into molds containing spherical balls of a material with a lower vaporization temperature. The powdered metal can be converted into porous metal by high-temperature treatment so the material in the spheres is vaporized leaving connected pores of predetermined size and distribution.[14]

In 1973 M. T. Karagianes of the Battelle Pacific Northwest Laboratories (Richland, Wash.) reported on investigations made under the sponsorship of the Atomic Energy Commission.

> Porous metals, if they are to function properly in skeletal replacement devices, must meet such criteria as biocompatibility, corrosion resistance, strength, flexibility in fabrication, tissue adherence, and permanent fixation.
>
> At the beginning of the seventies Battelle-Northwest developed Void Metal Composite (VMC), based on unalloyed titanium,

> a metal having controlled porosity . . . which, when implanted in bone, allows tissue ingrowth resulting in hard bone union between implant and skeleton. . . . Once it was established that tissue ingrowth and calcification did occur, we switched to a titanium alloy VMC containing 6% aluminum and 4% vanadium. This greatly increased strength without altering cellular ingrowth characteristics. . . . Even though tissue ingrowth produces a strong bone/metal union in many cases, certain practical complications can frustrate the successful application of this principle to the surgical solution of orthopedic and orthodontic problems. One of the most serious complications arises if movement of the implant is permitted during the healing period. . . . One must also keep in mind, as orthopedic surgeons are quick to suggest, that successful porous implant creates somewhat of a surgical problem should it, for some reason, have to be removed at a later time. . . . Exhaustive studies remain to be carried out prior to clinical trials; however, such continued work is well justified by the short-term results obtained to date by different investigators.[15]

David Williams of the University of Liverpool questions the continued use of metals as implant material, pointing out that an implant which is ignored by the tissues can never become fully incorporated into the body. "Without a degree of incorporation, the implant must be considered unstable, and, given certain conditions, could move." He concludes that implants of the future are likely to be made of ceramics or degradable polymers rather than "inert" metals and plastics and will have porous rather than smooth surfaces. He foresees that the prosthesis of the future will be designed "to stimulate complete regeneration of the tissue that it has replaced, making itself redundant in the process."[16]

John G. Suelzer, consultant to the Indianapolis Airport Authority, has raised the interesting question of the reaction of airport metal detectors to orthopedic devices. He points out that, while almost all external braces will trip the machines (causing no real problem since their existence can be readily demonstrated), the reaction to implanted devices is largely unpredictable. Therefore, he suggests that "people who have implanted metallic devices be supplied by their physician with evidence that an implant does exist." [17]

Pacemakers

Electrical currents were used to induce contraction of the heart muscle as early as the eighteenth century, but the concept of the modern pacemaker dates from 1932 when Albert Solomon Hyman of New York produced a cumbersome machine designed to achieve resuscitation from cardiac arrest. Twenty years later Paul M. Zoll of the Harvard Medical School, who with William Kouwenhoven of Johns Hopkins, another heart pioneer, would receive the Albert Lasker research award in November 1973, successfully treated heart block by repetitive electrical stimuli applied to the chest wall. However, while external pacemakers operating through surface electrodes gained immediate acceptance, the high voltages involved produced painful side effects that rendered repeated use undesirable. At about the same time, however, the advent of open heart surgery brought a need to avoid the development of inadvertent, and often lethal, heart block in the course of the operation. The answer lay in the introduction of wires sutured to the myocardium and brought out through the chest wall. The current required in this approach was not enough to produce pain.

The next step was to apply to patients with chronic heart block what had been developed to meet an acute situation. Small battery-powered pacemakers with silicone rubber-insulated bipolar electrodes were introduced, but long-term use was inhibited by the risk of infection occurring where the electrodes entered the skin and the danger of accidental disconnection of the externally worn apparatus. The first successful self-contained pacemaker carrying its own power supply was introduced in 1960.

> In the ensuing decade, increasingly sophisticated instrumentation has been developed. Implanted pacemaker systems operating through transvenously placed endocardiac electrodes became popular in 1963, and soon thereafter implantable devices were introduced, the output of which was programmed from spontaneous atrial or ventricular electrical activity. A recent report of the clinical use of a nuclear energy-powered pacemaker marked the introduction of an implantable power source with an extremely long life.[18]

The essential elements in a pacemaker are a pulse generator and the electrode(s). The former has ordinarily been powered by zinc-mercuric oxide batteries and housed in titanium, which shields against radiated electromagnetic fields. The electrode has a solid platinum core, a platinum-iridium tip, and leads of intertwined platinum ribbon wound around Dacron cores.

Theoretically the zinc-mercuric oxide battery should operate effectively for four or five years; experience has shown that it cannot be expected to last more than two or three years. It is obviously unsatisfactory to cut open a patient frequently to replace a costly, still functioning pacemaker with spent batteries by a new one with new batteries. The answer may lie in radioisotopes.

Isotope choice for pacemakers is limited by technology to plutonium-238, promethium-147, and thulium-171. The use of the last has so far been prohibited by cost and by purification problems. Pm-147 is attractive because of its low cost and adaptability to beta-excited direct energy conversion, but the cost of Pu-238 is competitive, and it has a long half life and a minimal effect on surrounding tissues (low dose rate). (It is anticipated that a pacemaker powered by plutonium-238 will continue to operate for ten years.)

In May 1970 the first nuclear-energy-powered pacemaker was implanted in a sixty-eight-year-old Frenchwoman; in July 1972 William M. Chardack and Andrew Gage of the Veterans Administration Hospital, Buffalo, New York, implanted the first two such pacemakers in the United States. Someday there will be pacemakers powered by the body's electrical impulses that will never have to be recharged.

The introduction of metal detectors at airports raised the question whether exposure to electromagnetic fields might have dangerous effects on passengers with permanently implanted pacemakers. Tests undertaken by the Potomac Fund for Cardiovascular Research and the Federal Aviation Administration had reassuring results:

> It seems that fears for the safety of patients carrying implanted pacemakers can be allayed. No effect was produced by the weapons detectors in patients with standard unipolar pacemakers. The majority of patients with permanent pacemakers will fall into one of the unaffected groups, i.e., those with fixed-rate ventricular pacers or unipolar triggered ventricular pacemakers. . . .
>
> All the pacemakers tested have built-in mechanisms designed to protect the patient against extraneous electrical interference, and these operated successfully in the pacemakers that were affected by the magnetic field.
>
> In every instance in which there was a rate alteration, the rate remained within safe limits.[19]

Of much more recent development than the cardiac pacemaker is the brain pacemaker of Irving S. Cooper, the effectiveness of which was announced to the profession in September 1973.

> Victims of intractable epilepsy and hypertonia [abnormal tension of muscles] have a murky existence, fraught with medical helplessness and personal despair. Equally gloomy—in appearance—are the old buildings housing St. Barnabas Hospital for Chronic Diseases in the Bronx, New York. But inside that hospital, the sun shines a couple of times a week as neurosurgeon Irving S. Cooper, MD, PhD, only 50 years old, adds another bead—possibly the largest yet—to his string of achievements in "functional neurosurgery."
>
> As of mid-summer, the majority of more than 30 patients with intractable epilepsy, cerebral palsy, or spasticity due to stroke have experienced marked improvement after the implantation of tiny stimulatory electrodes on the cortex of the cerebellum.[20]

The Cooper brain pacemaker involves one or more silicon-coated Dacron mesh plates to which four or eight pairs of platinum-disc bipolar electrodes are attached. By surgery that Cooper has described as "not particularly stressful," the electrodes are implanted in the back of the head directly on the tiny areas of the brain that regulate certain body movements. The electrode plates are connected by wires under the skin to a small radio receiver that is implanted in the patient's chest before brain surgery begins. The receiver gets its signal, by way of a small antenna taped to the patient's chest directly over the receiver, from a transmitter about the size of a package of cigarettes carried in a pocket or purse.

Cooper emphasizes that the brain pacemaker is not a cure, but implants have helped patients incapacitated by the effects of severe epilepsy, cerebral palsy, and spasticity due to stroke.

The first epilepsy patient to undergo brain stimulation had been hit on the head by a baseball four years earlier. Drug therapy had accomplished little, but after ten months of stimulation he was virtually seizure-free and appeared to be a perfectly normal twenty-four-year-old.

A thirteen-year-old boy with cerebral palsy could not even sit up. Eight months after surgery he could feed himself and swim in a pool, although he still could not walk.

An eighteen-year-old, who had had seizures most of his life and had been totally incapacitated for eight years, received a brain pacemaker in February 1973. With medication decreased, both his psychomotor and grand mal seizures stopped, along with his uncontrollable aggressive behavior toward himself and others. In May a wire of one of the stimulators broke and he suffered an almost complete relapse. Repair of the break reestablished control and by September he was working as a machinist.

Cooper believes the step he has taken will "unquestionably lead to a whole new application of physiological surgery of the brain. It will prove that some diseases are reversible and will stimulate research. In turn research will increase knowledge about the cerebellum and lead to new therapies for neurological diseases." [21]

Artificial Limbs

"Amputation is not an operation of destruction but is rather one of construction to form a new organ of locomotion and so to make the patient mobile again." [22]

There is evidence that amputation dates back to the final period of the Stone Age, but it was not until Roman times that attempts were made to replace severed limbs, particularly hands. An iron hand is, for example, mentioned by Pliny. Iron hands show up again in the fifteenth century. Still in existence is an iron hand fabricated for Götz von Berlichingen (1480–1562). With his right hand replaced, he continued warring for almost forty years. Ambroise Paré included in his *Oeuvres* (published in 1575) illustrations of artificial hands, arms, and legs. While the iron hand he depicted looked highly mechanical, even sophisticated, "his aim was primarily a cosmetic one, and his prostheses were no more self-activating than, say, artificial teeth." [23]

In the evolution of the modern artificial limb emphasis has been on self-activation. Granted that "the best such limb is a most inadequate substitute," especially since "we simply do not know how to 'plug in' to the central nervous system, other than in the crudest and most empirical manner" and therefore cannot achieve "control over the electrical signals from the brain itself," [24] there has been continuing progress in the development of self-powered prostheses.

While the need for limb replacement does not exclusively derive from amputation (the thalidomide tragedy of the late fifties and early sixties cannot be overlooked), in amputation cases careful surgery and nursing go a long way toward preparing the stump for the fitting of a prosthesis. This became of increased importance when, after World War II, "sockets of laminated plastic molded over plaster-of-Paris casts of a man's stump, were made to replace those that were hand carved and hand worked." [25]

The original powered artificial arms harnessed some body movement (such as a shrug of the shoulder) to provide the wearer with control over at least one operation of his prosthesis. Early in this century a more sophisticated approach attempted to use the muscles of the stump in essentially their original role by surgically modifying the muscle so that when flexed it would pull upon a control cable.

> The advantage of "cineplasty," as the operation was known, was that it permitted a more natural action to be used to move the artificial limb. The drawbacks proved first to be the difficulty of keeping the new tunnel clean and free from infection; and also that it meant a surgical modification of the body that was perhaps ahead of its time, and which both the patient and those close to him seemed to find hard to accept. For these two reasons the operation is little practised today.[26]

The next turn was toward limbs powered by extrinsic devices—carbon dioxide stored under pressure, electric batteries, and so on—but "power packs, whether of bottled gas or electric cells, are shortlived and too heavy; gas- and electric-powered motors respond crudely compared with our muscles; mechanical joints are clumsy and unadaptable." [27]

In the early 1950s Norbert Wiener (1894–1964) of the Massachusetts Institute of Technology suggested the possibility of using myoelectric currents (signals up to 25 millivolts generated by muscle fiber when it contracts), picked up by a sensitive electrode pressed against the skin over the muscle, to control the motions of a prosthesis. By the mid-fifties research teams led by Alfred Nightingale of London and Kobrinsky in Russia had developed simple forearm prostheses that utilized this procedure. By 1966 over 1,000 myoelectric artificial arms had been fitted in Russia compared with a handful in Britain.

With understandable emphasis on developing prostheses that might begin to substitute for our astonishingly versatile natural limbs, scant attention has been focused on the materials used in their construction. Three elements were essential—strength, lightness of weight, and electric conductivity. While an artificial leg, easily concealed by a trouser leg, might be constructed entirely of a lightweight metal (such as aluminum punctured with circular holes to reduce weight), in the arm and hand appearance must be taken into consideration. Obviously an arm that revealed to the eye aluminum tube rods with wires running among them from stainless steel electrodes and naked aluminum elbows and wrists could be embarrassing to the wearer and disturbing to the observer. Therefore, the "arm" is enclosed in a plastic laminated shell.

The main question involves the hand. "Cosmetic" hands are available, but they do not have the grip-strength of the metal split-hook "hand" that has become a familiar sight. The solution has been found in interchangeable hands. The hook, used for work, or in the privacy of the wearer's home, that has a three-point grip of up to ten pounds, can be unplugged and exchanged for the cosmetic hand in which the thumb and first two fingers are "active," moving toward each other in a caliper motion and providing a maximum fingertip force of six pounds. But there is more to this than providing a psychologically necessary cosmetic hand in substitution for the work hand. "On this principle, there is no reason why an amputee should not have a host of clip-on power tools, all of which he could operate with split-second accuracy. With suitable attachments he could drill, solder, weld, screw, bolt, polish—in fact, perform a variety of tasks with greater skill and reliability than a handed worker." [28]

One footnote to the versatility of the prosthetic hand. Until very recently the wearer of a hook prosthesis could not tell how much pressure he was exerting on anything he might pick up or use. The result could be breakage of or damage to delicate items he might be required to handle. In September 1973 Frank Clippinger, Jr., an orthopedist at Duke University Medical Center, reported the development of a hook wired to a surgically implanted electrical stimulator, connected directly to the medial or main arm nerve, which registers a mild tingling, the intensity of which tells the wearer how much force he is exerting.

> The four patients who have already been fitted with the experimental prosthesis report that it feels and performs much more like a normal hand than earlier devices. Milton Williamson, 47, an elementary school principal, used to wear clip-on bow ties; using Clippinger's arm he can now tie his own. [29]

16

Nuclear Medicine

The twentieth-century decline in the use of metals both in therapy and as repair materials has been offset by the introduction of metals in a form perhaps more precious than what has gone before—as radioisotopes.

On 8 November 1895 Wilhelm Konrad Röntgen (1845–1923), then professor of physics at the German University of Würzburg, was working late in a dark laboratory. He was experimenting with an early form of vacuum discharge tube devised by the English physicist and chemist, William Crookes (1832–1919), for the study of cathode rays. He had made the tube lightproof by covering it with black paper. When its beam accidentally hit on some crystals of barium platinocyanide lying on the table, the crystals began to fluoresce. Röntgen's next step was to direct his "rays" on a platinobarium screen set up nine feet away. Opaque objects placed in front of the screen produced varying degrees of intensity. When he placed his hand in front of the screen, he saw dense shadows of the bones with outlines of the flesh. When the screen was replaced by a photographic plate, a permanent picture was obtained. (The original screen, modified and improved, has been perpetuated in fluoroscopy.)

Surgeons in Berlin and Vienna were quick to recognize the diagnostic potential of Röntgen's "X rays," and their expectations were amply realized in the Greek-Turkish War (1897) and the Spanish-American War (1898) when X rays were used to diagnose fractures and locate bullets. X-ray therapy started with the discovery in 1896 by Leopold Freund (1868–1944) of Vienna of the effectiveness of radiation in the removal of excessive growths of hair. The next five years were given over to empirical therapy of skin lesions, notably lupus, psoriasis, favus, eczema, superficial tumors, and baldness.

A new period began in 1900 when the Austrians Guido Holzknecht (1872–1931) and Robert Kienböck introduced scientific dosage. In 1902 Henrich Ernst Albers-Schönberg (1865–1921) of Hamburg, Germany, invented a compression diaphragm that intensified the image by cutting out secondary rays. He also showed the injurious effect of the rays on internal organs and devised the leaden chamber to protect operators from sterility. In 1903 Georg Perthes (1869–1927) of Leipzig introduced deep therapy, and at about the same time Nicholas Senn (1844–1909), the Swiss-born professor of surgery at Rush Medical College, Chicago, applied X rays to the treatment of leukemia.

Inspired by Röntgen's discovery, the French physicist Henri Becquerel (1852–1908) decided to find out whether X rays were emitted by fluorescent bodies under the action of light. He chose salts of uranium for his experiment. The outcome was not what he had anticipated, but he learned that uranium salts *not* exposed to light spontaneously emitted rays of an unknown nature. A compound of uranium that had been in darkness for several months made an impression on a photographic plate from which it was separated by a sheet of black paper. Investigating further, he found that pitchblende, the chief source of uranium, was far more radioactive than could be accounted for by its uranium content. Becquerel concluded that pitchblende must also contain an unknown element with the power of affecting a photographic plate. This was in 1896. Becquerel suggested that his friend Marie Curie (1867–1934) attack the problem, an undertaking in which she was soon joined by her husband, Pierre Curie (1859–1906). Radium was discovered in 1898 and isolated in 1902.[1]

By one account the recognition of radium as a therapeutic agent was due to accident. Becquerel, carrying a fragment of uranium in his waistcoat pocket, suffered a burn similar to that produced by X rays. This suggested the use of radium in conditions for which X rays had proven therapeutic effect. Specifically, radium therapy was applied to lupus in 1901 and malignant tumors in 1903. Ultrapenetrating radium radiation, that was potent for neoplasms without damaging healthy tissues, was largely developed between 1906 and 1919 by Henri Dominici (1867–1919), an Englishman of Corsican descent.

At the beginning of the twentieth century it was accepted among physicists and chemists that the atom was indestructible and unchangeable—an atom of oxygen remained an atom of oxygen. Not so, said New Zealand-born, English-trained Ernest Rutherford (1871–1937) and English-born Frederick Soddy (1877–1956), at least in the case of radioactive atoms. In the early days of the present century Rutherford arrived at the concept, still subscribed to by physicists, that the atom consists of a central positively charged nucleus and outer negatively charged electrons.

By 1912 many scientists were on the track of isotopes—identical elements that differ only in their number of neutrons (the uncharged particles in the nucleus of the atom). Their disovery is generally credited to the Polish-American chemist Kasimir Fajans.

In 1918 Rutherford achieved the objective of the medieval alchemists if not their specific goal. An atom of nitrogen bombarded by an alpha particle was converted into an atom of oxygen plus a leftover proton.

> Not only was it possible to convert one familiar element like nitrogen into another like oxygen; bombardment also produced previously unknown species of matter—isotopes of the familiar elements, having the same number of protons and electrons but different atomic weights. The oxygen produced by Rutherford's original experiment, for example, was not ordinary oxygen-16, with an atomic weight of 16, but rather a heavier isotope, oxygen-17. A whole new world, it seemed, might be hidden in the interior of the atom.[2]

Radioisotopes in Therapy

Radioisotope treatment involves the concentrating in malfunctioning tissue of an amount of radiation that will destroy the malfunctioning tissue without damage to healthy tissue.

After cancer, one of the early candidates for radioactive attack was hyperthyroidism—a condition resulting from overworked thyroid glands. X-ray treatment of this problem was common practice in the 1920s—until it became apparent that the rays, as then handled, were destroying healthy as well as defective tissue. In the mid-forties iodine-131, produced cheaply at the Institute of Nuclear Sciences, Oak Ridge, Tennessee, was substituted for X rays in radioactive treatment of hyperthyroidism. (It is to be noted that radioisotopes employed in therapy and diagnosis are not necessarily isotopes of metals.) A further setback to radioactive therapy occurred in the fifties when the public joined the government in a radiation (fallout) scare. Doctors were loath to recommend radiation therapy except in hopeless cases of cancer. But the radioiodine treatment of hyperthyroidism was continued, as well as treatment of leukemia with phosphorus-32, introduced in 1936 by hematologist John H. Lawrence of the University of California at Berkeley. Phosphorus-32 also became and has remained the treatment of choice in polycythemia vera (excess of red blood cells) and the intracavity treatment of pleural and peritoneal effusions in patients with diagnosed cancer.

By the mid-forties some dissatisfaction had arisen in respect of the use of phosphorus-32 in the treatment of acute and chronic leukemias and malignant lymphomas. Between 1944 and 1956 Paul F. Hahn and his co-workers at Vanderbilt and Meharry experimented with radioactive colloids intravenously administered. Radioiodine and manganese-52 were tried first, but the final choice fell on gold-198 which, among other advantages, can be administered in a single dose to ambulatory patients, obviating hospitalization and the need for repeated visits. Gold-198 is also employed in cases of accumulation of fluid in the pleural or peritoneal cavity and in a variety of other therapies. In 1962 Joseph Greenberg of the Long Island Jewish Hospital, New York, suggested that yttrium-90 was of palliative use in the treatment of multiple myeloma (formation of tumor masses in bone marrow). Cesium-137 is employed in cancer therapy and its particular application to cancer of the breast has been detailed by Robert Amalric and Jean-Maurice Spitalier of the *Centre Régional de Lutte contre le Cancer à Marseille* in a book published in 1973.[3] Californium-252 (produced by bombarding curium-242 with alpha particles) is proving of value in the treatment of intestinal and intracavity cancer, largely because it requires a lesser amount of oxygen during the period of irradiation than is the case with other radioisotopes.

It would seem that the use of radioactive isotopes should be an ideal way to irradiate areas of the skin and accessible mucous membranes. One could select the isotope that emitted beta rays of the energy required to give exactly the maximum penetration of tissue desired. The applicator could be compact and portable and less expensive than the x-ray machines used in contact therapy. Irregular and relatively inaccessible lesions could be "covered" and normal tissue spared. Radium and radon have been used for this purpose in dermatology since the earliest days of radiotherapy. However, these have obvious disadvantages. The applicator cannot be "tailored" to fit the lesion and the radiation is never pure beta, a gamma ray component always being present. Adequate protection is more difficult and, if radon is used, the escape of the gas is an ever present hazard.[4]

A solution offered in 1946 was simplicity itself. Pieces of blotting paper were soaked in a phosphorus-32 solution and applied to the tissue to be treated. But this method was found to have disadvantages. Twenty years later skin lesions and superficial malignancies, scattered over a wide area, were being treated with strontium-90 used in conjunction with yttrium-90, providing a higher emission of beta particles. Actually this approach was not new. The use of a strontium-90–yttrium-90 beta particle applicator in ophthalmology had been suggested in 1950. "The advantages of this type of radiation over x rays for superficial lesions of the eye are numerous because no radiation is delivered to the interior of the globe, and the lens, which is particularly susceptible to injury, is spared. The area to be treated can be sharply circumscribed and treated by direct contact so that large doses can be delivered."[5] There are today many commercially available strontium applicators specially designed for various purposes and sites of application.

Radioisotopes in Diagnosis

The application of radioisotopes to diagnosis lagged seriously behind their use in therapy. In fact their full diagnostic potential was not recognized until around 1950, and another decade passed before adequate tracing equipment was developed. This is rather surprising when one considers that the Hungarian-born physicochemist Georg von Hevesy (1885–1966) had conducted experiments that had led him by 1923 to the establishment of principles that have remained basic to the use of radioisotopes as tracers.

Before radioisotopes, the doctor who wished to investigate the functioning of an internal organ had four choices: he could feel with his hands (palpation); he could explore endoscopically; he could employ chemical tests in some cases; he could turn to X rays.

Some organs cannot be X-rayed at all. Those that can show as a one-dimensional shadow on a film. Even with bones, the prime field for X-ray examination, cancer of the bone is not recognizable until it has become extensive. Not only do radioisotopes reveal change in the activity of the bone cells quite early, but internal organs—such as the liver, thyroid, lung, spleen, kidney, heart, and pancreas—can be observed *in function.* Such observation is of first importance in determining whether an organ is healthy, with its size and shape offering contributory evidence. When a radioisotope is introduced into a healthy organ, the organ will retain it or dispose of it. If retention or disposal is greater or less than the established norm, the organ is not functioning properly. The feasibility of radioisotope diagnosis rests on the fact that certain radioisotopes introduced into a patient's body find their way to certain organs. An appropriate radioisotope is introduced and the organ is scanned to see what happens when the radioisotope gets to it.

Today observations are made with the rectilinear scanner and the scintillation camera. The first "scanner" was that of Benjamin Cassen of the University of California at Los Angeles. In 1949 he moved a gamma ray detector across a thyroid gland to observe the activity of iodine-131 that had been introduced.

In the ensuing quarter century scanners have been improved and refined in innumerable ways. Of more recent development was the camera. It offers at least two advantages over the scanner. It produces an image in a fraction of a second in contrast with the scanner's fraction of an hour. The camera presents a picture in motion, the scanner a still picture.

The list of radioisotopes in diagnostic use today is a long one, including a number of nonmetals; the survey that follows covers only radioisotopes of metals.

Beginning with the major organs, mercury-197, mercury-203, technetium-99m, ytterbium-169, indium-113, strontium-85, and strontium-87m are employed in brain scanning, with copper-64, arsenic-72, arsenic-74, gallium-68, and potassium-42 as additional aids to brain tumor localization. (Until very recently brain scans were performed within one hour after the intravenous introduction of technetium-99m, but evidence has been educed that delayed scanning—twenty-four hours—was more effective. "Delayed scanning proved to be superior because of the increased field obtained in most intracranial lesions. The delayed technetium scan should be the procedure of first choice. Delayed mercury [-203] scanning should be used when delayed technetium scans are negative or equivocal in a patient suspected of having a lesion adjacent to the base of the skull, especially in the subtentorial region." [6]) Lung visualizations involve indium-113m and technetium-99m, and mercury-197 and -203 are used to detect pulmonary infarction. Cardiac scanning employs potassium-43 and cesium-129, and myocardial blood flow is checked with cesium-131, cesium 134m, rubidium-84 and rubidium-86. Blood flow problems are investigated with chromium-51, sodium-22, and sodium-24, and the first of these radioisotopes is used in blood volume determination and red blood count labeling. Liver scans are made with gold-198, technetium-99m, and indium-113m, while tumors in the liver are located with chromium-51 and molybdenum-99.

Masses in the pancreas are detected with selenium-75. Kidney function scans and tumor localizations involve mercury-197, mercury-203, technetium-99m, indium-113m, and ytterbium-169. Chromium-51, mercury-197, technetium-99m, and indium-113m are the radioisotopes for spleen visualization. Soft tissue tumors, including colonic and rectal tumors, are localized with gallium-67. Breast tumors are detected with potassium-42. Bone scans on patients with diagnosed cancer involve strontium-85 which, along with strontium-87m, calcium-47, gallium-72, and, most recently, thulium-167, has also been employed in localizing bone tumors. Bone marrow imaging employs indium-111, and tumors in the bone marrow are located with gold-198, iron-52, gallium-68, and technetium-99m. Gastrointestinal protein loss is studied with chromium-51; cerebrospinal fluid flow is checked with indium-111; pernicious anemia is diagnosed with cobalt-57, -58, and -60.

The major constituent of the human body is water, about two-thirds of the total body water being intracellular, one-third extracellular, with sodium and chloride predominating in the extracellular "space," potassium and phosphate in the intracellular. But the compartments are not closed to each other, and constant interchange occurs. Among the radioisotopes employed in determining that proper balance is being maintained are sodium-22, sodium-24, potassium-42, and rubidium-84.

Iron turnover studies involve iron-52, -55, and -59. Magnesium absorption is measured with magnesium-28, calcium absorption with calcium-45 and -47, and intestinal absorption studies employ cobalt-58 and -60. Copper metabolism is checked with copper-64.

Measuring the rate of passage of food through the stomach can be of diagnostic importance but previous tests, including the use of barium capsules and microspheres labeled with chromium-51, cesium-129, and technetium-99m, have proved unsatisfactory both in point of accuracy and of practicability. In January 1973 a group of Canadian doctors announced the effectiveness of indium-113m used in conjunction with a 10-crystal rectilinear scanner. "The technique is simple, the results reproducible and it has proved ideal for assessing the results of drug therapy in patients with gastric retention. . . . The technique should be easily adaptable to a gamma camera with region-of-interest capability." [7]

Antepartum bleeding, especially during the third trimester of pregnancy, presents the physician with a major diagnostic problem in attempting accurately to establish the position of the placenta as a guide to subsequent management. Diagnosis based on digital examination was contraindicated for a variety of reasons.

> Any method which allows this diagnosis to be made reliably without the necessity for vaginal examination must be considered advantageous. The most commonly used procedures in this regard are those involving radiographic and isotopic studies. . . . Radiologic determination of the placental site by soft tissue, contrast, or displacement techniques has about a 95 per cent accuracy rate.
>
> In general, isotopic methods of determing placental localization are based on the concept that maternal blood flows slowly through the placenta and pools in the placental sinusoids. When the maternal blood cells are tagged with a suitable radioactive isotope location of the placenta will become manifest as a local accumulation of the tracer, which can be outlined by surveying the abdomen with an appropriate detector.

The radioisotope selected for this purpose was chromium-51 and the procedure was recommended because the radiation exposure of both mother and fetus was low.[8] Indium-113m has also been used in placental imaging.

The very number of radioisotopes employed in medicine today might suggest that nuclear medicine is close to providing a see-all and a cure-all, but the discipline is still highly experimental and complexities challenge the clinical use of radioactive materials. The uncertainties involved are well illustrated by what the *Journal of the American Medical Association* described as the "Off-and-on ruling on technetium use." Reference was to the fact that the Atomic Energy Commission banned the use of technetium-99m as a lung-scanning agent in May 1973 and reinstated it the following September. The ban was based on three deaths that followed administration of the radioisotope in preparation for lung scans between July and September 1972. (They did not come to the AEC's attention until the following May and there had been no further deaths in the interim.)

What prompted the AEC to react as it did?

> Deaths due to lung scanning are rare—only three had been reported in American literature up to that time. None of these had been associated with technetium-ferrous hydroxide particles. . . .
>
> There were other good reasons to believe the scanning agent probably did not contribute to the patients' deaths. Two of the patients were 86 years old, and gravely ill with suspected pulmonary thromboemboli. The third, a 50-year-old man, also was critically ill with bilateral bronchopneumonia. His autopsy revealed advanced atherosclerotic disease.[9]

The conclusion to be drawn is that nuclear medicine and those required to "police" it are still at a nervous stage.

But the use of radioisotopes in therapy and diagnosis unquestionably has an expanding future in which the precious metals of medicine will play their part.

Reference Notes
Selected Bibliography
Illustration Credits
Index

Reference Notes

1. Out of the Crucible

1. Bruce A. Rogers, *The Nature of Metals* (Ames: Iowa State University Press, 2d ed., 1964), p. 4.
2. J. Gordon Parr, *Man, Metals and Modern Magic* (Ames: Iowa State College Press, 1958), pp. 6–8.
3. *Ibid.*, p. 7.
4. *Ibid.*, p. 135.
5. *Ibid.*, p. 230.

2. What Makes a Metal Precious?

1. *Encyclopedia Britannica,* 14th ed. (1965), s.v. "Metal."
2. Parr, *Man, Metals and Modern Magic,* p. 51.
3. Republic of the Philippines, Department of Health, Culion Sanitarium, *Culion 1906–1956,* ed. H. W. Wade (Manila: Bureau of Printing, 1956), pp. 11–12.
4. Parr, *op. cit.*, p. 54.
5. Geoffrey Marks and William K. Beatty, *The Medical Garden* (Scribners, 1971), pp. 31, 33.
6. George Edward Trease, *Pharmacy in History* (London: Baillière, Tindall and Cox, 1964), p. 235.

3. The Early History of Metals in Medicine

1. Exod. 32:20.
2. Quoted in W. T. Fernie, *Precious Stones: for Curative Wear; and Other Remedial Uses: Likewise the Nobler Metals* (Bristol, England: John Wright, 1907), p. 362.
3. Henry E. Sigerist, *A History of Medicine, vol. I, Primitive and Archaic Medicine* (New York: Oxford University Press, 1951), p. 322.
4. *Ibid.*, pp. 283, 322, 471–72.
5. A. C. Wootton, *Chronicles of Pharmacy,* 2 vols. (London: Macmillan, 1910), I:40.
6. Sigerist, *op. cit.*, p. 486.
7. Quoted in Fernie, *op. cit.*, pp. 363–64.
8. Quoted in Obed Simon Johnson, *A Study of Chinese Alchemy* (Shanghai: Commercial Press, 1928), pp. 59, 63.
9. *Ibid.*, p. 59.
10. C. J. S. Thompson, *The Mystery and Romance of Alchemy and Pharmacy* (London: Scientific Press, 1897), p. 8.
11. *Ibid.*, p. 41.
12. Wootton, *op. cit.*, I:90.
13. Fielding H. Garrison, *An Introduction to the History of Medicine* (Philadelphia: W. B. Saunders, 4th ed., 1929), p. 112.
14. Thompson, *op. cit.*, p. 46.
15. Fernie, *op. cit.*, pp. 368–69.
16. Cecilia C. Mettler, *History of Medicine,* ed. Fred A. Mettler (Philadelphia: Blakiston, 1947), p. 184.
17. Joseph Walsh, "Galen Visits the Dead Sea and the Copper Mines of Cyprus," *Bulletin of the Geographic Society of Philadelphia* XXV (1927): 99.
18. Garrison, *op. cit.*, pp. 28–30.
19. Quoted in John Stewart Milne, *Surgical Instruments in Greek and Roman Times* (Oxford: Clarendon, 1907), p. 154.
20. Quoted *ibid.*, p. 134.
21. John Scarborough, *Roman Medicine* (London: Thames and Hudson, 1969), p. 83.
22. Milne, *op. cit.*, p. 24.
23. *Ibid.*, pp. 26–27, Plate V.
24. Quoted *ibid.*, p. 49.
25. *Ibid.*, p. 51.
26. Celsus quoted *ibid.*, p. 53.

27. Quoted *ibid.*, p. 100.
28. Quoted *ibid.*, pp. 101–2.
29. Paul quoted *ibid.*, p. 112.
30. *Ibid.*, p. 121.
31. Quoted in Scarborough, *op. cit.*, p. 84.
32. *Ibid.*, p. 86.

4. The Role of the Alchemists

1. H. Stanley Redgrove, *Alchemy: Ancient and Modern* (London: William Rider, 2d ed., 1922), pp. 1–2.
2. M. M. Pattison Muir, *The Story of Alchemy and the Beginnings of Chemistry* (London: G. Newnes, 1902), pp. 105–6.
3. J. Schouten, *The Rod and the Serpent of Asklepios: Symbol of Medicine* (Amsterdam: Elsevier, 1967), pp. 120–21.
4. William Jerome Wilson, "Traditional Background of Greco-Egyptian Alchemy," *Ciba Symposia* III, no. 5 (August 1941): 927–28.
5. Garrison, *An Introduction to the History of Medicine*, p. 55n.
6. Wilson, *op. cit.*, p. 934.
7. Ernst von Meyer, *A History of Chemistry from the Earliest Times to the Present Day, being also an Introduction to the Study of the Science* (New York: Macmillan, 3d English ed. from the 3d German ed., 1906), p. 31.
8. F. P. Venable, *A Short History of Chemistry* (Boston: Heath, 1896), p. 13.
9. Johnson, *A Study of Chinese Alchemy*, p. 69.
10. Mettler, *History of Medicine*, p. 190.
11. Garrison, *op. cit.*, p. 137.
12. Redgrove, *op. cit.*, p. 45.
13. Meyer, *op. cit.*, p. 35.
14. [Roger Bacon], *The Mirror of Alchimy, Composed by the thrice-famous and learned Fryer,* Roger Bachon, *sometimes fellow of* Martin *Colledge; and afterwards of* Brasen-nose *Colledge of Oxenforde* (London: Printed for Richard Olive, 1597), pp. 1–2.
15. Thompson, *The Mystery and Romance of Alchemy and Pharmacy*, pp. 54–55.
16. Quoted in Redgrove, *op. cit.*, p. 49.
17. Peter Bonus, *The New Pearl of Great Price*, trans. A. E. Waite, quoted in Redgrove, *op. cit.*, p. 50.
18. Garrison, *op. cit.*, p. 204.
19. Quoted in Anna M. Stoddart, *The Life of Paracelsus, Theophrastus von Hohenheim,* 1493–1541 (London: William Rider, 1915), p. 43.
20. Henry E. Sigerist, "Paracelsus after Four Hundred Years," in *Henry E.*

Sigerist on the History of Medicine, ed. Félix Martí-Ibáñez (New York: MD Publications, 1960), p. 164.
21. Stoddart, *op. cit.,* pp. 32–33.
22. Quoted *ibid.,* p. 31.
23. Quoted *ibid.,* p. 32.
24. Peter Ramus (1515–1572), quoted *ibid.,* p. 45.

5. The Triumphant Chariot of Antimony

1. Trease, *Pharmacy in History,* p. 142.
2. Wootton, *Chronicles of Pharmacy,* I:228.
3. *Ibid.,* p. 229.
4. *Basil Valentine His Triumphant Chariot of Antimony, with annotations of Theodore Kirkringius, M.D.* (London: Printed for Dorman Newman, 1678), pp. 116–17.
5. Garrison, *An Introduction to the History of Medicine,* p. 205n; Wootton, *op. cit.,* I:225.
6. Wootton, *op. cit.,* I:243–44.
7. *Supra,* note 4, pp. 33–34.
8. *Ibid.,* p. 51.
9. *Ibid.,* p. 52.
10. *Ibid.,* pp. 54–55.
11. *Ibid.,* p. 59.
12. Wootton, *op. cit.,* I:381.
13. John Huxham, *Medical and Chemical Observations upon Antimony* (London: Printed for John Hinton, 1756), pp. 5–6.
14. P. O'Connell, "Composition and Therapeutic Uses of True James' Powder," *Chicago Medical Journal and Examiner* XXIX (1897):163–64.
15. Huxham, *op. cit.,* p. 6.
16. *Ibid.,* pp. 75–76, 78.
17. *Ibid.,* pp. 73–74.
18. William Blizard, "Experiments and Observations of the External Use of Emetic Tartar," *London Medical Journal* VIII (1787):57–58.
19. *Ibid.,* pp. 58–59.

6. The Liquid Metal

1. Wootton, *Chronicles of Pharmacy,* I:410.
2. John W. Francis, "Observations on the Natural and Medical History of Mercury," *The American Medical Recorder* V, no. 3 (July 1822):401.

3. Quoted *ibid.*, p. 406.
4. Leonard J. Goldwater, *Mercury; A History of Quicksilver* (Baltimore: York Press, 1972), p. 110.
5. *Ibid.*, p. 113.
6. *Ibid.*, pp. 117–18.

7. The Noble Metals

1. Mettler, *History of Medicine*, p. 205.
2. Wootton, *Chronicles of Pharmacy*, I:392.
3. *Ibid.*, pp. 393–94.
4. Mettler, *op. cit.*, pp. 204–5.
5. Wootton, *op. cit.*, I:395.
6. John C. Cheesman, *An Inaugural Dissertation on the Medical Properties of Gold* (New York: Printed for Collins and Co., 1812), pp. 14–16.
7. *Ibid.*, pp. 7, 16–17.
8. *Ibid.*, pp. 3, 17–19, 28.
9. Wootton, *loc. cit.*
10. Samuel R. Percy, "Lectures on New Remedies and their Therapeutic Applications. Lecture III. Aurum.—Gold," *American Medical Times* IV, no. 6 (8 February 1862):75.
11. Richard Powell, "Observations on the internal use of *Nitrate* of *Silver*, in certain convulsive Affections," *Medical Transactions of the Royal College of Physicians* IV (1813):88.
12. *Ibid.*, pp. 88–89.
13. *Ibid.*, p. 91.
14. *Ibid.*, pp. 100–101.
15. *Ibid.*, p. 91.
16. John Higginbottom, *An Essay on the Application of Lunar Caustic in the Cure of Certain Wounds and Ulcers* (London: Longman, 1826).
17. Stephen Brown, "Observations upon the use of Nitrate of Silver, as a remedy in various local affections," *American Medical Recorder* XIII (1828): 117.
18. John Higginbottom, "Using the Nitrate of Silver in the Cure of Inflammation, Wounds, and Ulcers," *Lancet* I (1850):74–75.
19. *Ibid.*, p. 74.
20. *Ibid.*, p. 75.
21. *Ibid.*, p. 77.
22. William Pepper, "On the Internal Administration of Nitrate of Silver, and on the Occurrence of a Blue Line upon the Gums as the Earliest Sign

of Argyria," *Transactions of the College of Physicians of Philadelphia,* 3d ser., III (1876–77):39.

23. *Ibid.,* pp. 40–41.
24. C. H. B. Lane, "Observations on the Oxide of Silver; and an Abstract of Cases in which it has been administered," *Medico-Chirugical Review,* XXXIII (1840):296.
25. Francis Bennett, "Uterine Hæmorrhage Treated by Oxide of Silver," *Lancet* I (1850):77–78.
26. Francis R. Packard, "The History of Some Famous Quacks," *Bulletin of Johns Hopkins Hospital* XV (1904):321–23; Joseph T. Smith, "Historical Sketch of Dr. Elisha Perkins, Inventor of the Metal Tractors," *Maryland Medical Journal* LIII (1910):173.
27. Jacques M. Quen, "Elisha Perkins, Physician, Nostrum-Vendor, or Charlatan?" *Bulletin of the History of Medicine* XXXVII (1963):159.
28. *Ibid.,* p. 160.
29. Quoted *ibid.,* p. 162.
30. *Ibid.,* pp. 165–66

8. The Semimetals

1. Wootton, *Chronicles of Pharmacy,* I:387.
2. Samuel A. Moore, *An Inaugural Dissertation on the Values of White Oxide of Bismuth* (New York: Printed by T. & J. Swords, 1810), pp. 13–14, 17, 35.
3. Alex. Marcet, "Observations on the Medical Use of the White Oxyd of Bismuth," *Memoirs of the Medical Society of London* VI (1805):155–56.
4. *Ibid.,* pp. 156–57.
5. *Ibid.,* p. 170.
6. *Ibid.,* p. 171n.
7. Quoted in Moore, *op. cit.,* pp. 38–39.
8. I. P. Garvin, "Remarks on the use of the Sub-Nitrate of Bismuth in certain gastric derangements," *Southern Medical and Surgical Journal,* n.s. I (July 1845):353, 357–58.
9. *Ibid.,* p. 357.
10. Joh. Hermann Baas, *Outlines of the History of Medicine and the Medical Profession,* trans. H. E. Handerson (New York: J. H. Vail, 1889), p. 720.
11. Nathaniel Potter, "An Essay on the Medical Properties and Deleterious Qualities of Arsenic," in Charles Caldwell, ed., *Medical Theses selected from among the Inaugural Dissertations . . . of the University of Pennsylvania, and*

of other medical schools in the United States (Philadelphia: Thomas and William Bradford, 1805), *Preface* [p. 47].

12. William Withering, *The Miscellaneous Tracts* (London: Longman, Hurst, 1822), II:472–74.
13. E. S., "Arsenic as a Remedy," *The Boston Medical and Surgical Journal* LI (1854):37; Potter, *op. cit.*, p. 62.
14. Quoted in E. S., *loc. cit.*
15. Potter, *op. cit.*, pp. 68–69.
16. Benjamin Rush, *Medical Inquiries and Observations*, 2d U.S. ed. (Philadelphia: Thomas Dobson, 1794), I:235.
17. Potter, *op. cit.*, p. 50.
18. Mettler, *History of Medicine*, p. 651.
19. Anthony Todd Thomson, "On the Preparation and Nature of the Iodide of Arsenic; with observations and experiments on its medicinal and poisonous properties," *Lancet* I (1838–1839): pp. 176, 182.
20. Thomas Hunt, "Memoir on the Medicinal Action of Arsenic," *Transactions of the Provincial Medical and Surgical Association, London* XVI (1849):384n.
21. *Ibid.*, pp. 393, 409–10, 412–13.
22. James Begbie, "Arsenic: its Physiological and Therapeutical Effects," *Edinburgh Medical Journal* III (9 May 1858):961.
23. Stewart Lockie, "On the Use of Arsenic as a Blood and Cardiac Tonic," *The British Medical Journal* II (7 December 1878):828–29.
24. F. A. H. LaRue, "An Arsenic-Eater," *The Boston Medical and Surgical Journal* LXXIV (1866):439, 441–42.
25. Iretus Greene Cardner, "Studies in Materia Medica. Arsenic in Consumption, and generally in Diseases where there is a tendency to Death by Asthenia; its Physiological and Psychological Action," *New York Medical Journal* IX (1869):132.
26. Horatio R. Storer, "Upon the Arsenical Atmosphere and Arsenical Hot Spring of the Solfatara at Pozzuoli (near Naples), in the Treatment of Consumptives," *Lancet* II (29 September 1877):457.

9. The Common Metals

1. Wootton, *Chronicles of Pharmacy*, I, 426.
2. Ting Kai Li, "The Functional Role of Zinc in Metalloenzymes," in *Zinc Metabolism*, ed. Ananda S. Prasad (Springfield, Ill.: Charles C Thomas, 1966), p. 48.

3. Wootton, *op. cit.*, I:427.
4. Campbell De Morgan, "On the Use of the Chloride of Zinc in Surgical Operations and Injuries, and especially, in Operations for the Removal of Cancerous Tumours," *British and Foreign Medico-Chirurgical Review* XXXVII (1866):202.
5. *Ibid.*, p. 206.
6. *Ibid.*
7. *Ibid.*, p. 209.
8. *Ibid.*, p. 210.
9. *Ibid.*, p. 214.
10. Joseph Lister, "On the Antiseptic Principle in the Practice of Surgery," *Lancet* I (1867):95–96.
11. Campbell De Morgan, "On the use of Chloride of Zinc solution in the treatment of Abscess connected with diseased Joints," *Transactions of the Clinical Society of London* I (1868):138.
12. *Ibid.*, p. 139.
13. *Ibid.*, p. 140.
14. *Ibid.*, p. 142.
15. Arthur H. Nicols, "Fatal Poisoning by Chloride of Zinc Applied to an Epithelioma of the Lip: with Remarks," *Boston Medical and Surgical Journal* CXV, no. 15 (14 October 1886):345.
16. *Ibid.*, pp. 343–44.
17. *Ibid.*, p. 347.
18. "Navy Medical Reports. No VI. Extracts from Official Reports upon the Effects of Chloride of Zinc in Deodorizing Offensive Effluvia from Cesspools, Sewers, etc., and in Decomposing Poisonous Emanations from the Bodies of those Affected by Contagious Diseases," *Medical Times & Gazette of London*, n.s. VII (1853):341.
19. *Ibid.*, p. 344.
20. From an advertising piece distributed by E. S. Horne & Co., Philadelphia (1879).
21. Jerry K. Aikawa, *The role of Magnesium in Biologic Processes* (Springfield, Ill.: Charles C Thomas, 1963), p. 3.
22. John H. Steel, *An Analysis of the Congress Spring, with Practical Remarks on its Medical Properties,* revised and corrected by John L. Perry (New York, 1861), pp. 1, 12.
23. Jon B. Eklund and Audrey B. Davis, "Joseph Black Matriculates: Medicine and Magnesia Alba," *Journal of the History of Medicine and Allied Sciences* XXVII, no. 4 (October 1972):410.

24. *Ibid.*, p. 417.
25. *Ibid.*, pp. 413–14.
26. Joseph Black, "*Experiments upon* Magnesia alba, *Quicklime and some other Alcaline Substances,*" *Essays and Observations, Physical and Literary (Edinburgh)* II (1770):176.
27. Humphry Davy, "Electro-Chemical Researches, on the Decomposition of the Earths; with Observations on the Metals obtained from the alkaline Earths, and on the Amalgam procured from Ammonia," *Philosophical Transactions of the Royal Society of London* XCVIII (1808):339–41.
28. *Ibid.*, p. 342.
29. *Ibid.*, p. 345.
30. *Ibid.*, p. 346.
31. James Henry, "An Account of an Improved Mode of Administering Sulphate of Magnesia (Epsom Salt), whereby it is rendered an agreeable, safe, and efficaceous purgative, applicable to almost every case in which a purgative is required," *Edinburgh Medical and Surgical Journal* XLI (1834):48–49, 54.
32. Wootton, *op. cit.*, I:401–2.
33. *Ibid.*, pp. 399–400.
34. John Cule, "The iron mixture of Dr. Griffith," *Pharmaceutical Journal* CXCVIII, no 5398 (15 April 1967):399–400.
35. Quoted *ibid.*, p. 400.
36. "A Study in the Action of Iron," *Medical Communications, Massachusetts Medical Society* XIII (1882):74.
37. *Ibid.*, p. 75.
38. T. C. Adam, "On the Remedial Powers of the Persesquinitrate of Iron," *American Journal of the Medical Sciences* XXIV (1839):61, 64.
39. Isaac Casselberry, "On the Use of Iron," *American Journal of the Medical Sciences* n.s. XXXV (1858):330, 332, 335.
40. E. N. Chapman, "The History, Preparation and Therapeutical Uses of the Citro-Ammoniacal Pyrophosphate of Iron, named in brief Pyrophosphate of Iron," *Boston Medical and Surgical Journal* LXVI, no. 1 (6 February 1862):12, 20.
41. Abraham Livezey, "Observations on the Use of Iron in Obstetricy," *Boston Medical and Surgical Journal* LIII (1855):446.
42. J. Milner Fothergill, "When Not to Give Iron," *Practitioner* XIX (1877):183; R. W. Crighton, "Note on the Administration of Iron in Pulmonary Phthisis and Senile Anæmia," *Practitioner* XX (1878):11.
43. Wootton, *op. cit.*, I:364–65.

44. [Thomas] Goulard, *A Treatise on the Effects and Various Preparations of Lead, particularly of the Extract of Saturn, for Different Chirurgical Disorders,* trans. G. Arnaud (London: Printed for P. Elmsly, 1775), pp. 1–3.
45. *Ibid.,* pp. 196, 198, 201, 203–216.
46. *Ibid.,* pp. 11, 30, 52, 59, 63–64, 68, 73, 89–90, 96, 98–108, 116–17, 125, 132–33, 135–36, 143–46, 157–58, 165.
47. William Laidlaw, "Remarks on the Internal Exhibition of the Acetate of Lead, chiefly with the View of determining to what Extent it may be safely administered in the Cure of Diseases, especially in Uterine Hæmorrhages," *Boston Medical and Surgical Journal* I (1828–29):147.
48. A. Kimbal, "Observations on the use of large doses of Acetate of Lead in some of the diseases of Alabama and Georgia," *Western Journal of Medicine and Surgery (Louisville),* 2nd ser., I (1840):218–223.
49. R. H. Goolden, "On the Use of Sulphate of Manganese in Various Diseases," *London Medical Gazette* XXXV (1844–45):646.
50. E. H. Davis, "The Salts of Manganese," *American Medical Monthly* II (1854):343–44.
51. *Ibid.,* p. 344.
52. Franklin H. Martin, "Manganese as an Emmenagogue," *Chicago Medical Journal and Examiner* L (February 1855):119–20, 126–27.
53. R. Leaman, "Some Clinical Observations on the Therapeutic Uses of Bromide of Nickel," *Medical News, Philadelphia* XLVI (18 April 1885):427.
54. *Ibid.,* pp. 427–29.

10. The Earth Metals

1. Adair Crawford, "On the Medicinal Properties of the Muriated Barytes," *Medical Communications (Society for Promoting Medical Knowledge) London* II (1790):301–2, 346–48.
2. *Ibid.,* p. 359.
3. J. Warburton Begbie, "The Therapeutic Actions of Muriate of Lime," *Edinburgh Medical Journal* XVIII (July 1872): 47–48.
4. James Wood, "Observations on the Efficacy of Muriate of Lime, in the Cures of Scrofula and the other States of Debility," *Edinburgh Medical and Surgical Journal* I (1805):148–49.
5. James Sanders, *Treatise on Pulmonary Consumption in which a New View of the Principles of its Treatment is supported by Original Observations on every Period of the Disease, etc., etc.* (Edinburgh: Walker & Greig, 1808), p. 112.

6. John Thomson, *Lectures on Inflammation* (Edinburgh: W. Blackwood, 1813), p. 196.
7. Quoted in Begbie, *op. cit.*, pp. 51–52.
8. Quoted *ibid.*, p. 52.
9. *Ibid.*
10. *Ibid.*, pp. 46–47.
11. John Cleland, "On the Use of Saccharated Lime in Medicine," *Edinburgh Medical Journal* V (1859–1860):113–15.
12. John Cleland, "The Use of Saccharated Lime in Typhus Fever and Other Complaints, *Practitioner* XV (December 1875):401.
13. R. Blacke, "On the Use of Lacto-Phosphate of Lime as an Analeptic Medicament in Adynamic Fevers and in Convalescence," *Practitioner* VIII (February 1872):65–66, 69.
14. H. W. Jones, "Observations on the Therapeutic Action of the Oxalate of Cerium in the Vomiting of Pregnancy," *Chicago Medical Journal* n.s. IV (February 1861):65, 69; Charles Lee, "On the Therapeutic Use of the Oxalate of Cerium," *American Journal of the Medical Sciences* n.s. XL (October 1860):391, 393.

11. The Alkali Metals

1. John Butter, "On the Effects of Nitre," *Edinburgh Medical and Surgical Journal* XIV (January 1818):34–35, 38–39.
2. G. H. E. Zimmerman in *Journal der practischen Heilkunde* (1843), quoted in J. B. Brown, "On the Medical properties of Nitrate of Soda," *Northwestern Medical and Surgical Journal* n.s. XI (1854):62.
3. *Ibid.*
4. *Ibid.*, pp. 62, 64.
5. H. Macnaughton Jones, "Nitrate of Potash and Quinine as Febrifuges," *British Medical Journal* I (1 March 1873):224.
6. William Channing, "On the Iodo-Hydrargyrate of Potassium; its Chemical History and Therapeutic Uses," *American Journal of the Medical Sciences* XIII (1833–34):391, 398, 401–3.
7. H. W. Berg, "A Plea for the Substitution of the Sodium Iodide for the Potassium Iodide in Therapeutics," *Archives of Medicine* (*New York*) XI (1884):145,151,155–56.
8. Alexander Russell Simpson, "On the Use of Chlorate of Potassa in Obstetrics and Gynecology," *Transactions of the American Gynecological Society* XIII (1888):415.
9. *Ibid.*, pp. 418–19.

10. E. J. Fountain, "Report of Cases of Phthisis, Scrofula, and other Diseases treated by Chlorate of Potash, with Remarks on its Mode of Administration and the Importance of using a Preparation free from Impurity," *American Medical Monthly* XV (February 1861):97.
11. Trail Green, "The New Official Chlorate," *Journal of the American Medical Association* III (July 1884):60.
12. Garrison, *An Introduction to the History of Medicine,* p. 844; [Karl] Binz, "The Therapeutic Employment of Bromide of Potassium," *Practitioner* XII (1874):6.
13. Quoted in Binz, *op. cit.,* pp. 7–8.
14. James Begbie, "Notice of some of the Therapeutic Effects of the Bromide of Potassium," *Edinburgh Medical Journal* XII (1866):481.
15. *Ibid.,* pp. 482–84, 492.
16. "Case in which Bromide of Potassium was Pushed to its Full Extent," *British Medical Journal* II (1869):415.
17. Lunsford P. Yandell, Jr., "Remarks on the Therapeutic Value of Bromide of Potassium," *Western Journal of Medicine* IV (1869):592.
18. [Francis Edmund] Anstie, "The English Stand-Point respecting the Value of Bromide of Potassium," *Practitioner* XII (1874):19.
19. Binz, *op. cit.,* p. 14.
20. Quoted in Anstie, *loc. cit.*
21. *Ibid,* pp. 20–21.
22. *Ibid.,* pp. 22–25, 28.
23. T. J. Hudson, "Bromide and Iodide of Sodium; their Therapeutic Advantages over Bromide and Iodide of Potassium," *Lancet* II (22 December 1883):1081.
24. *Ibid.,* p. 1082.
25. Henry M. Field, "The Superior Value of Bromide of Sodium," *Boston Medical and Surgical Journal* CVIII (10 May 1883):438–39.
26. *Ibid.,* p. 439.
27. William Stephenson, "On the Action and Uses of Phosphate of Soda in Small Doses," *Edinburgh Medical Journal* XIII (1867–68):336–37, 343–44.
28. Henry G. Piffard, "Salt in Dermal Hygiene and Therapeutics," *Transactions of the American Dermatological Association* XI (1887):34–37, 40.
29. George Henry Fox, "The Dermatological Value of Sulpholeate of Sodium," *Journal of Cutaneous and Genito-Urinary Diseases* VIII (1890):169–71.
30. G. A. Baxter, "Silicate of Soda—Some New Methods of Use in Surgery," *Transactions of the Southern Surgical and Gynecological Association* III (1890):295–99.

31. "Salicylate of Lithia in Rheumatism," *Medical News* XLVIII (9 January 1886):36.

12. Metals in Diagnosis and Internal Medicine

1. For a generalized discussion of the achievements of medicine *see* Geoffrey Marks and William K. Beatty, *The Story of Medicine in America* (New York: Scribners, 1973).
2. "How Copper Colors the Tissues' Life," *Medical World News* XIII (14 April 1972):61.
3. Ananda S. Prasad, "Zinc Deficiency Syndrome in Man: A Historical View," in Carl C. Pfeiffer, ed., *International Review of Neurobiology Supplement 1* (New York: Academic Press, 1972), p. 1.
4. *Ibid.*, p. 3.
5. R. J. Clayton, "Double-blind Trial of Oral Zinc Sulphate in Patients with Leg Ulcers," *British Journal of Clinical Practice* XXVI (August 1972):368–70.
6. T. Hallböök and E. Lanner, "Serum-Zinc and Healing of Venous Leg Ulcers," *Lancet* II (1972):780–82.
7. "Serum Zinc and Healing Venous Leg Ulcers," *CMD* XL (July 1973):551–52.
8. Aba Marshak and Gabriel Marshak, "Zinc Sulfate Therapy for Vocal Cord Granulomas," *Journal of Laryngology and Otology* LXXXVII (June 1973):575.
9. "Just How Zinc-deficient Are Americans?" *Medical World News* XIII (4 August 1972):63.
10. John H. Menkes, "Kinky Hair Disease," *Pediatrics* L, no. 2 (August 1972):181.
11. "A New Genetical Disorder of Copper Metabolism," *Medical Journal of Australia* II, no. 6 (5 August 1972):291–92.
12. *Ibid.*, p. 292.
13. I. Herbert Scheinberg and Irmin Sternlieb, "Metabolism of Trace Metals," in Philip K. Bondy, ed., *Duncan's Diseases of Metabolism,* 6th ed. (Philadelphia: W. B. Saunders, 1969), p. 1325.
14. Carl C. Pfeiffer and Venelin Iliev, "A Study of Zinc Deficiency and Copper Excess in Schizophrenias," in Pfeiffer, *International Review of Neurobiology Supplement 1,* p. 160.
15. Scheinberg and Sternlieb, *loc. cit.*
16. Jack E. Ansell and Munsey S. Wheby, "Pica: Its Relation to Iron Deficiency," *Virginia Medical Monthly* XCIX (September 1972):951–53.

17. Scheinberg and Sternlieb, *op. cit.*, p. 1329.
18. "A chromium key to glucose tolerance," *Medical World News,* XIII (19 May 1972):46–47.
19. I. J. T. Davies, *The Clinical Significance of the Essential Biological Metals* (London: Heinemann, 1972), p. 100.
20. M. L. Scott, "The Selenium Dilemma," *Journal of Nutrition* CIII (June 1973): 808.
21. "Trace Elements Take Spotlight at Giant Biology Conference," *Medical World News* XIV (11 May 1973): 37.
22. James P. Knochel, "Exertional Rhabdomyolysis," *New England Journal of Medicine* CCLXXXVII, no. 18 (2 November 1972): 928.
23. James D. Gowans and Mohammad Salami, "Response of Rheumatoid Arthritis with Leukopenia to Gold Salts," *New England Journal of Medicine* CCLXXXVIII, no. 19 (10 May 1973): 1007–8.
24. John C. Krantz, Jr., "Aluminum Salts in Medicine," *CMD* XXXIX, no. 9 (September 1972): 894.

13. Metals in Other Therapies

1. John R. Cochran, "A New Ointment Containing Zinc Peroxide," *Northwestern University Medical School Quarterly Bulletin* XVIII (Spring 1944): 41–43.
2. Arthur T. Risbrook, et al., "Gold Leaf in the Treatment of Leg Ulcers," *Journal of the American Geriatrics Society* XXI (July 1973): 325, 329.
3. Carl A. Moyer, "The Treatment of Severe Thermal Injury," *Western Journal of Surgery, Obstetrics and Gynecology* LXII (1954):40.
4. *Ibid.*
5. *Ibid.*, pp. 45–46.
6. Carl A. Moyer et al., "Treatment of Large Human Burns with 0.5% Silver Nitrate Solution," *Archives of Surgery* XC (June 1965):816–19.
7. John A. Moncrief, "Burns," *New England Journal of Medicine* CCLXXXVIII, no. 9 (1 March 1973):447.
8. I. Kelman Cohen et al., "Hypogeusia, Anorexia, and Altered Zinc Metabolism Following Thermal Burns," *Journal of the American Medical Association* CCXXIII, no. 8 (19 February 1973):914–16.
9. "Serum Zinc and Healing of Venous Leg Ulcers," *CMD* XL (July 1973):552.
10. For the status of the mentally ill in the eighteenth and nineteenth centuries, see Geoffrey Marks and William K. Beatty, *The Story of Medicine in America,* pp. 64–67; *Women in White* (New York: Scribners, 1972), pp. 175–87.

11. Ronald R. Fieve, "Lithium in Psychiatry," *International Journal of Psychiatry* IX (1970–71):383.
12. *Ibid.*, pp. 384, 386, 391.
13. *Ibid.*, pp. 402–3.
14. Phillip Polatin, "Lithium Carbonate Prophylaxis in Affective Disorders," *Diseases of the Nervous System* XXXIII (July 1972):472–73.
15. *Ibid.*, pp. 473–75.
16. Burton P. Grimes, "Lithium," *Minnesota Medicine* LV (September 1972): 832.
17. David Samuel and Zehava Gottesfeld, "Lithium, Manic-Depression, and the Chemistry of the Brain," *Endeavour* XXXII, no. 117 (September 1973):127–28.
18. Roy G. Fitzgerald, "Mania as a Message," *American Journal of Psychotherapy* XXVI, no. 4 (October 1972):547, 553.
19. Card C. Pfeiffer and Venelin Iliev, "A Study of Zinc Deficiency and Copper Excess in the Schizophrenias," *International Review of Neurobiology, Supplement 1*, p. 142.
20. *Ibid.*, pp. 141, 162–63.

14. Additional Applications of Metals to Modern Medicine

1. [William H. Moss], "Bilateral Partial Vasectomy," *CMD* XXXIX (October 1972): 1068–70.
2. Robert T. Bliss, "A Vasectomy Prosthesis," *Illinois Medical Journal* CXLIII (May 1973): 430.
3. Daniel J. Preston and Charles F. Richards, "Use of Wire Mesh Prostheses in the Treatment of Hernia," *Surgical Clinics of North America* LIII, no. 3 (June 1973): 549, 553–54.
4. Francis Mitchell-Heggs and H. Guy Radcliffe Drew, *The Instruments of Surgery* (London: Heinemann, 1963), p. 92.
5. M. Richard Schoor, "Needles, Some Points to Think About," *Anesthesia and Analgesia (Cleveland)* XLV (1966):512.
6. *Ibid.*, p. 514.
7. S. B. Cheng and L. K. Ding, "Practical Applications of Acupuncture Analgesia," *Nature* CCXLII (17 April 1973): 559–60.
8. William W. Scott, "The Development of the Cystoscope," *Investigative Urology* VI, no. 6 (May 1969): 657.
9. *Ibid.*, p. 658.
10. E[dmund] Andrews, "The Interior of the Urethra Viewed by a Magnesium Light," *Chicago Medical Examiner* VIII (1867): 211–12.

11. Joseph H. Kiefer, "Edmund Andrews—Forgotten Pioneer of Chicago Urology," *Proceedings of the Institute of Medicine of Chicago* XXIX, no. 4 (July 1972): 131.
12. Scott, *loc. cit.*
13. David M. Wallace, "New Lamps for Old," *Proceedings of the Royal Society of Medicine* LXVI (May 1973): 458.
14. J. E. Sjöstedt, "The Vacuum Extractor and Forceps in Obstetrics: A Clinical Study," *Acta Obstetricia et Gynecologica Scandinavica,* Supp. 10 to XLVI (1967): p. 11.
15. J. A. Chalmers, "James Young Simpson and the 'Suction-Tractor,' "*Journal of Obstetrics and Gynæcology of the British Commonwealth* LXX (1963): 95–96.
16. Sjöstedt, *loc cit.*
17. Chalmers, *op cit.,* p. 96.
18. Sjöstedt, *op cit.,* pp. 11–12.
19. Chalmers, *op cit.,* p. 98.
20. Sjöstedt, *op. cit.,* pp. 16–21.
21. Harry Foreman, "Copper Seven Intrauterine Device," *Minnesota Medicine* LVI (June 1973): 474.
22. Lise Fortier et al., "Canadian Experience with a Copper-covered Intrauterine Contraceptive Device," *American Journal of Obstetrics and Gynecology* CXV, no. 3 (1 February 1973): 296.
23. John Newton et al., "Intrauterine Contraception Using the Copper-Seven Device," *Lancet* II, no. 7784 (4 November 1972): 951; Foreman, *op. cit.,* p. 475.
24. "IUDs May Wipe Out Gonorrhoea," *New Scientist* LVII (18 January 1973): 118.
25. Naomi Bluestone, "Otosclerosis," *New York State Journal of Medicine* LXXII (1972): 2001.
26. Francis A. Sooy et al., "Stability of Hearing Over an Eight-Year Period Following Wire-Vein Stapedectomy for Otosclerosis," *Annals of Otology, Rhinology, and Laryngology* LXXXII, no. 1 (January–February 1973): 13, 16.
27. Hugh Beckman and H. Saul Sugar, "Neodymium Laser Cyclocoagulation," *Archives of Ophthalmology* XC, no. 1 (July 1973): 27–28.
28. Leon Goldman et al., "High-Power Neodymium-YAG Laser Surgery," *Acta Dermatovenereologica (Stockholm)* LIII, no. 1 (1973): 45, 48–49.
29. William Meyerhoff et al., "Gold Foil Closure of Oroantral Fistulas," *Laryngoscope* LXXXIII (June 1973): 940–43.
30. Fletcher C. Derrick, Jr., "Gold Leaf in Vesicovaginal and Vesicorectal Fistulas," *Journal of Urology* CX, no. 3 (September 1973): 296.

31. James T. McClowry and Maynard Ferguson, "Space Age Process Applied to Medical Problem," *Pennsylvania Medicine* LXXVI (March 1973): 41–42.

15. Bones, Joints, Implants, and Prostheses

1. Charles Orville Bechtol et al., *Metals and Engineering in Bone and Joint Surgery* (Baltimore: Williams and Wilkins, 1959) p. v.
2. W. J. Peters, "Implant Materials and Their Use in Orthopedic Surgery," *Canadian Journal of Surgery* XVI, no. 3 (May 1973): 179.
3. Quoted *ibid.*
4. Quoted in Bechtol, *op. cit.*, pp. 4–5.
5. Thomas Annandale, "On the Use of Steel Pins in the Practice of Surgery," *Scottish Medical and Surgical Journal* I (1897): 868–69.
6. Peters, *op. cit.*, p. 181.
7. "Learning to Replace Knees with Metal, Plastic Devices," *Journal of the American Medical Association* CCXXIII, no. 10 (5 March 1973): 1085.
8. Bernard L. Manale et al., "Total Hip Replacement," *Journal of the Kentucky Medical Association* LXXI (July 1973): 440, 443.
9. John H. Arnett, "Delayed Infections after Total Hip Replacement," *Journal of the American Medical Association* CCXXIII, no. 9 (26 February 1973): 1042–43; Marks and Beatty, *The Story of Medicine in America,* p. 338; Cyril P. Monty and A. K. Sahukar, "Experience with Stanmore Metal-to-plastic Hip Prosthesis," *Proceedings of the Royal Society of Medicine* LXVI (June 1973): 512.
10. *Supra,* note 7.
11. *Clinical Orthopaedics and Related Research,* no. 94 (July–August 1973), p. 2.
12. G. Blundell Jones, "Total Knee Replacement—The Walldius Hinge," *ibid.*, pp. 50, 57.
13. *Drug Research Reports,* XVI, no. 42 (17 October 1973): RN-2.
14. Robert F. Rushmer, *Medical Engineering* (New York: Academic Press, 1972), pp. 301–2.
15. M. J. Karagianes, "Porous Metals as a Hard Tissue Substitute," *Biomaterials, Medical Devices, and Artificial Organs* I, no. 1 (June 1973): 171–73, 175–77, 179.
16. David Williams, "Step Nearer the Redundant Implant," *New Scientist* LVIII, no. 843 (28 April 1973): 221–23.
17. John G. Suelzer, "Orthopedic Devices and Airport Metal Detectors," *Journal of the Indiana State Medical Association* LXVI (June 1973): 380–81.

18. William M. Chardack, "Cardiac Pacemakers," in *Davis-Christopher Textbook of Surgery,* 10th ed., edited by David C. Sabiston, Jr. (Philadelphia: W. B. Saunders, 1972), II:2080.
19. John M. Keshishian et al., "The Behavior of Triggered Unipolar Pacemakers in Active Magnetic Fields," *Journal of Thoracic and Cardiovascular Surgery* LXIV, no. 5 (November 1972): 777–78.
20. Gail McBride, "Cerebellar Stimulation Aids Victims of Intractable Hypertonia, Epilepsy," *Journal of the American Medical Association* CCXXV, no. 12 (17 September 1973): 1441.
21. *Ibid.,* p. 1446.
22. M[iroslaw] Vitali, "Rehabilitation of the Amputee," *Proceedings of the Royal Society of Medicine* LIX (January 1966): 3.
23. David Fishlock, *Man Modified* (London: Jonathan Cape, 1969), p. 83.
24. Donald Longmore, *Spare-Part Surgery* (New York: Doubleday, 1968), p. 61; Fishlock, *op. cit.,* p. 82.
25. Carl Peter Mason, "Design for Powered Prosthetic Arm System for the Above-Elbow Amputee," in *Veterans Administration Bulletin of Prosthetic Research* 10–18 (Fall 1972): 10.
26. Fishlock, *op. cit.,* p. 85.
27. Longmore, *op. cit.,* pp. 61, 63.
28. *Ibid.,* p. 75.
29. "Clippinger's Arm," *Time* (10 September 1973), p. 61.

16. Nuclear Medicine

1. For an account of the discovery and isolation of radium *see* Marks and Beatty, *Women in White,* pp. 202–8.
2. Ruth and Edward Brecher, *The Rays—a History of Radiology in the United States and Canada* (Baltimore: Williams and Wilkins, 1969), p. 332.
3. Robert Amalric and Jean-Maurice Spitalier, *Césiumthérapie Curative des Cancers du Sein* (Paris: Masson, 1973).
4. Solomon Silver, *Radioactive Nuclides in Medicine and Biology, Vol II, Medicine,* 3d ed. (Philadelphia: Lea & Febiger, 1968), p. 480.
5. *Ibid.,* p. 481.
6. Stephen B. Meisel et al., "Comparison of Early and Delayed Technetium and Mercury Brain Scanning," *Radiology* CIX, no. 1 (October 1973): 117.
7. Geoffrey Coates et al., "Measurement of the Rate of Stomach Emptying Using Indium-113m and a 10-Crystal Rectilinear Scanner," *Canadian Medical Association Journal* CVIII, no. 2 (20 January 1973): 180, 182.
8. Frederick A. Brodrick and Wayne A. Cotnoir, "Localization of the Pla-

cental Site with Chromium^{51}-Tagged Erythrocytes," *Rhode Island Medical Journal* LVI (June 1973): 226.

9. "Off-and-On Ruling on Technetium Use Illustrates a Problem in Medicine," *Journal of the American Medical Association* CCXXV, no. 10 (3 September 1973): 1165.

Selected Bibliography

"A Chromium Key to Glucose Tolerance." *Medical World News* XIII (19 May 1972): 46–47.

AIKAWA, JERRY K. *The role of Magnesium in Biologic Processes.* Springfield, Ill.: Charles C Thomas, 1963.

ANAN'YEV, M. G., ed. *New Soviet Surgical Apparatus and Instruments and their Application.* New York: Pergamon Press, 1961.

"An Essential Role for Nickel?" *Medical World News* XIII (21 April 1972): 69–70.

ANSELL, JACK E. and WHEBY, MUNSEY S. "Pica: Its Relation to Iron Deficiency." *Virginia Medical Monthly* XCIX (September 1972): 951–54.

ANSTIE, [FRANCIS EDMUND]. "The English Stand-Point Respecting the Value of Bromide of Potassium," *Practitioner* XII (1874): 19–28.

"A Study of the Action of Iron," *Medical Communications. Massachusetts Medical Society,* XIII (1882): 67–76.

Basil Valentine His Triumphant Chariot of Antimony, with Annotations of Theodore Kirkringius, M.D. with The True Book of the Learned Synesius a Greek Abbot taken out of the Emperour's Library, concerning the Philosopher's Stone. London: Printed for Dorman Newman, 1678.

BECHTOL, CHARLES ORVILLE et al. *Metals and Engineering in Bone and Joint Surgery.* Baltimore: Williams and Wilkins, 1959.

BELCHER, E. H. and VETTER, H., eds. *Radioisotopes in Medical Diagnosis.* New York: Appleton-Century-Crofts, 1971.

SELECTED BIBLIOGRAPHY

BERGERSEN, BETTY S. and KRUG, ELSIE E., in consultation with GOTH, ANDRES. *Pharmacology in Nursing.* 10th ed. St. Louis: C. V. Mosby, 1966.

BINZ, [KARL]. "The Therapeutic Employment of Bromide of Potassium." *Practitioner* XII (1874): 6–18.

BLACK, JOSEPH. *"Experiments upon* Magnesia alba, *Quicklime, and some other Alcaline Substances." Essays and Observations, Physical and Literary* (*Edinburgh*) II (1770): 172–248.

BLIZARD, WILLIAM. "Experiments and Observations on the external Use of Emetic Tartar." *London Medical Journal* VIII (1787): 57–60.

BLOOM, WILLIAM and FAWCETT, DON W. *A Textbook of Histology.* 9th ed. Philadelphia: Saunders, 1968.

BONDY, PHILIP K., ed. *Duncan's Diseases of Metabolism.* 6th ed. 2 vols. Philadelphia: Saunders, 1969.

BRECHER, RUTH and EDWARD. *The Rays—a History of Radiology in the United States and Canada.* Baltimore: Williams & Wilkins, 1969.

BROWN, STEPHEN. "Observations upon the use of the Nitrate of Silver, as a remedy in various local affections." *American Medical Recorder* XIII (1828): 116–27.

Bulletin of Prosthetics Research, 10–18 (Fall 1972). Veterans Administration, Washington, D.C.

BUTTER, JOHN. "On the Effects of Nitre." *Edinburgh Medical and Surgical Journal* XIV (January 1818): 34–39.

CHALMERS, J. A. "James Young Simpson and the 'Suction-Tractor.' " *Journal of Obstetrics and Gynaecology of the British Commonwealth* LXX (1963): 94–100.

CHEESMAN, JOHN C. *An Inaugural Dissertation on the Medical Properties of Gold.* New York: Printed by Collins and Co., 1812.

Clinical Orthopaedics and Related Research 94 (July–August 1973).

CRAWFORD, ADAIR. "On the Medicinal Properties of the Muriated Barytes." *Medical Communications* (*Society for Promoting Medical Knowledge*) *London* II (1790): 301–59.

CRENSHAW, A. H., ed. *Campbell's Operative Orthopaedics.* 2 vols. St. Louis: Mosby, 1971.

CULE, JOHN. "The iron mixture of Dr. Griffith." *Pharmaceutical Journal* CXCVIII, no. 5398 (15 April 1967): 399–401.

DAVIES, I. J. T. *The Clinical Significance of the Essential Biological Metals.* London: Heinemann, 1972.

DAVY, HUMPHRY. "Electro-Chemical Researches, on the Decomposition of the

Earths; with Observations on the Metals obtained from the alkaline Earths, and on the Amalgam procured from Ammonia." *Philosophical Transactions of the Royal Society of London* XCVIII (1808): 333–70.

DEBUS, ALLEN G. "The Paracelsians and the Chemists: the Chemical Dilemma in Renaissance Medicine." *Clio Medica* VII, no. 3 (1972): 185–99.

DIMOND, E. GREY. "Acupuncture Anesthesia—Western Medicine and Chinese Traditional Medicine." *Journal of the American Medical Association* CCXVII, no. 10 (6 December 1971): 1558–63.

DI PALMA, JOSEPH R., ed. *Drill's Pharmacology in Medicine.* 4th ed. New York: McGraw-Hill, 1971.

DRAKE, T. G. H. "Antique Pewter of Medical Interest." *Bulletin of the History of Medicine* X (1941): 272–87; XXIX (1955): 420–28.

EKLUND, JON B. and DAVIS, AUDREY B. "Joseph Black Matriculates: Medicine and Magnesia Alba," *Journal of the History of Medicine and Allied Sciences* XXVII, no. 4 (October 1972): 396–417.

FELLOWS, E. W. "The History of Some Surgical Instruments." *Australian and New Zealand General Practitioner* XXV (1954): 277–79, 308–13, 335–39.

FERNIE, W. T. *Precious Stones: for Curative Wear; and Other Remedial Uses: Likewise The Nobler Metals.* Bristol, England: John Wright, 1907.

FIELD, HENRY M. "The Superior Value of the Bromide of Sodium." *Boston Medical and Surgical Journal* CVIII (1883): 438–40.

FIEVE, RONALD R. "Lithium in Psychiatry." *International Journal of Psychiatry* IX (1970–71): 375–412.

FISHLOCK, DAVID. *Man Modified.* New York: Funk & Wagnalls, 1969.

FITZGERALD, ROY G. "Mania as a Message." *American Journal of Psychotherapy* XXVI, no. 4 (October 1972): 547–54.

FOREMAN, HARRY. "Copper Seven Intrauterine Device." *Minnesota Medicine* LVI (June 1973): 474–79.

FORTIER, LISE et al. "Canadian Experience with a Copper-Covered Intrauterine Device." *American Journal of Obstetrics and Gynecology* CXV, no. 3 (1 February 1973): 291–97.

FRANCIS, JOHN W. "Observations on the Natural and Medical History of Mercury." *American Recorder* V, no. 3 (July 1822): 395–407.

GARVIN, I. P. "Remarks on the use of the Sub-Nitrate of Bismuth in certain gastric derangements." *Southern Medical and Surgical Journal,* n.s. I, no. 7 (July 1845): 353–58.

GOLDWATER, LEONARD J. *Mercury—A History of Quicksilver.* Baltimore: York Press, 1972.

GOULARD, [THOMAS]. *A Treatise on the Effects and Various Preparations of Lead, particularly the Extract of Saturn, for Different Chirurgical Disorders.* A new edition, tr. by G. Arnaud. London: Printed for P. Elmsly, 1775.

HIGGINBOTTOM, JOHN. *An Essay on the Application of the Lunar Caustic in the Cure of Certain Wounds and Ulcers.* London: Longman, 1826.

———. "Using the Nitrate of Silver in the Cure of Inflammation, Wounds, and Ulcers." *Lancet* I (1850): 74–77.

HOOD, R. MAURICE et al. "The Use of Automatic Stapling Devices in Pulmonary Resection." *Annals of Thoracic Surgery* XVI, no. 1 (July 1973): 85–98.

"How copper colors the tissues' life." *Medical World News* XIII (14 April 1972): 61–62.

HULBERT, S. F. et al., eds. "Materials and Design Considerations for the Attachment of Prostheses to the Musculo-Skeletal System." *Journal of Biomedical Materials Research* VII, no. 3 (1973).

HUXHAM, JOHN. *Medical and Chemical Observations upon Antimony.* London: John Hinton, 1756.

JOHNSON, OBED SIMON. *A Study of Chinese Alchemy.* Shanghai: Commercial Press, 1928.

"Just How Zinc-deficient Are Americans?" *Medical World News* XIII (4 August 1972): 61, 63.

LANE, C. H. B. "Observations on the Oxide of Silver; and an Abstract of the Cases in which it has been administered." *The Medico-Chirurgical Review* XXXIII (1840): 289–96.

LAWALL, CHARLES H. *Four Thousand Years of Pharmacy.* Philadelphia and London: Lippincott, 1927.

LEAMAN, R. "Some Clinical Observations on the Therapeutic Uses of Bromide of Nickel." *Medical News, Philadelphia* XLVI (18 April 1885): 427–29.

LONGMORE, DONALD. *Spare-Part Surgery.* New York: Doubleday, 1968.

MARCET, ALEX. "Observations on the Medical Use of the White Oxyd of Bismuth." *Memoirs of the Medical Society of London* VI (1805): 155–173.

MARKS, GEOFFREY and BEATTY, WILLIAM K. *The Medical Garden.* New York: Scribners, 1971.

———. *The Story of Medicine in America.* New York: Scribners, 1973.

———. *Women in White.* New York: Scribners, 1972.

MARTIN, FRANKLIN H. "Manganese as an Emmenagogue." *Chicago Medical Journal and Examiner* L (February 1885): 119–27.

MAUGH, THOMAS H., II. "Trace Elements: A Growing Appreciation of Their Effects on Man." *Science* CLXXXI, no. 4096 (20 July 1973): 253–54.

MCCORD, CAREY P. and BAYLIS, SHIRLEY J. "The Origin of the Names of

Elemental Metals." *Journal of Occupational Medicine* XV (June 1973): 531–34.

MEYER, ERNST VON. *A History of Chemistry from Earliest Times to the Present Day, being also an Introduction to the Study of the Science.* Translated by George McGowan. 3d English ed. from the 3d German ed. New York: Macmillan, 1906.

MILNE, JOHN STEWART. *Surgical Instruments in Greek and Roman Times.* Oxford: The Clarendon Press, 1907.

MITCHELL-HEGGS, FRANCIS and DREW, H. GUY RADCLIFFE. *The Instruments of Surgery.* London: Heinemann, 1963.

MODELL, WALTER, ed. *Drugs of Choice 1970–1971.* St. Louis: Mosby, 1970.

MØLLER-CHRISTENSEN, VILHELM. *The History of the Forceps.* London: Oxford University Press, 1938.

MONCRIEF, JOHN A. "Burns." *The New England Journal of Medicine,* CCLXXXVIII, no. 9 (1 March 1973): 444–54.

MOORE, SAMUEL W. *An Inaugural Dissertation on the Medical Virtues of Bismuth; with Some Preliminary Observations on the Chemical Properties of That Metal.* New York: Printed by T. & J. Swords, 1810.

MORGAN, CAMPBELL DE. "On the Use of Chloride of Zinc Solution in the treatment of Abscess connected with diseased Joints." *Transactions of the Chemical Society of London* I (1868): 138–42.

———. "On the Use of the Chloride of Zinc in Surgical Operations and Injuries, and especially in Operations for the Removal of Cancerous Tumours." *British and Foreign Medico-Chirurgical Review* XXXVII (1866): 201–14.

MOYER, CARL A. et al. "Treatment of Large Human Burns with 0.5% Silver Nitrate Solution." *Archives of Surgery* XC (1965): 812–67.

MUIR, M. M. PATTISON. *The Story of Alchemy and the Beginnings of Chemistry.* London: G. Newnes, 1902.

"Navy Medical Reports. No VI. Extracts from Official Reports upon the Effects of Chloride of Zinc in Deodorizing Offensive Effluvia from Cesspools, Sewers, etc., and in Decomposing Poisonous Emanations from the Bodies of those Afflicted with Contagious Diseases." *Medical Times & Gazette of London,* n.s., VII (1853): 341–44.

New Surgical Equipment and Instruments and Experience in Their Use. Translated from the Russian by S. Shoshan and Y. Flancreich for the National Science Foundation, Washington, D.C., 1961.

NICHOLS, ARTHUR H. "Fatal Poisoning by Chloride of Zinc Applied to an Epithelioma of the Lip: with Remarks," *Boston Medical and Surgical Journal* CXV, no.15 (14 October 1886), 343–47.

O'CONNELL, P. "Composition and Therapeutic Uses of True James' Powder," *Chicago Medical Journal and Examiner* XXXIX (1879): 160–66.

OSOL, ARTHUR and PRATT, ROBERTSON, eds. *U.S. Dispensatory.* 27th ed. Philadelphia: Lippincott, 1973.

PARR, J. GORDON. *Man, Metals and Modern Magic.* Ames: Iowa State College Press, 1958.

PERCY, SAMUEL R. "Lectures on New Remedies and their Therapeutic Applications. Lecture III. Aurum.—Gold." *American Medical Times* IV, no. 6 (8 February 1862): 75–76.

PETERS, W. J. "Implant Materials and Their Use in Orthopedic Surgery." *Canadian Journal of Surgery* XVI, no. 3 (May 1973): 179–86.

PFEIFFER, CARL C., ed. *International Review of Neurobiology Supplement 1.* New York and London: Academic Press, 1972.

POLATIN, PHILLIP. "Lithium Carbonate Prophylaxis in Affective Disorders." *Diseases of the Nervous System* XXXIII (July 1972): 472–75.

POTCHEN, E. JAMES, ET AL. *Principles of Diagnostic Radiology.* New York: McGraw-Hill, 1971.

POTTER, NATHANIEL. "An Essay on the Medicinal Properties and Deleterious Qualities of Arsenic." In Caldwell, Charles, ed., *Medical Theses selected from among the Inaugural Dissertations . . . of the University of Pennsylvania, and of other medical schools in the United States.* Philadelphia: Thomas and William Bradford, 1805.

POWELL, RICHARD. "Observations on the internal use of *Nitrate* of *Silver,* in certain convulsive Affections." *Medical Transactions of the Royal College of Physicians* IV (1813): 85–102.

PRASAD, ANANDA S., ed. *Zinc Metabolism.* Springfield, Ill.: Charles C Thomas, 1966.

QUEN, JACQUES M. "Elisha Perkins, Physician, Nostrum-Vendor, or Charlatan?" *Bulletin of the History of Medicine* XXXVII (1963): 159–66.

REDGROVE, H. STANLEY. *Alchemy: Ancient and Modern.* 2d ed. London: Rider, 1922.

ROGERS, BRUCE A. *The Nature of Metals.* 2d. ed. Ames: Iowa State University Press, 1964.

SABISTON, DAVID C., ed. *Davis-Christopher Textbook of Surgery.* 10th ed. 2 vols. Philadelphia: Saunders, 1972.

SAMUEL, DAVID and GOTTESFELD, ZEHAVA. "Lithium, Manic-Depression, and the Chemistry of the Brain." *Endeavour* XXXII, no. 117 (September 1973): 122–28.

SCARBOROUGH, JOHN. *Roman Medicine.* London: Thames and Hudson, 1969.

SCHMIDT, J. E. *Medical Discoveries: Who & When.* Springfield, Ill.: Charles C Thomas, 1959.

SCHORR, M. RICHARD. "Needles, Some Points to Think About." *Anesthesia and Analgesia (Cleveland)* XLV, nos. 4 and 5 (1966): 509–526.

SCOTT, WILLIAM W. "The Development of the Cystoscope," *Investigative Urology* VI, no. 6 (May 1969): 657–661.

"Selenium plays hero, heavy, bit parts," *Medical World News* XIII (8 December 1972): 51–53.

SHUMAKER, WAYNE. *The Occult Sciences in the Renaissance.* Berkeley: University of California Press, 1972.

SIGERIST, HENRY E. *A History of Medicine. Vol. I, Primitive and Archaic Medicine.* New York: Oxford University Press, 1951.

SILVER, SOLOMON. *Radioactive Nuclides in Medicine & Biology. Vol. II, Medicine.* 3d ed. Philadelphia: Lea & Febiger, 1968.

SIMPSON, ALEXANDER RUSSELL. "On the Use of Chlorate of Potassa in Obstetrics and Gynecology." *Transactions of the American Gynecological Society* XIII (1888): 413–22.

SJÖSTEDT, J. E. "The Vacuum Extractor and Forceps in Obstetrics: A Clinical Study." *Acta Obstetricia et Gynecologica Scandinavica.* Supp. 10 to XLVI (1967).

STANLEY, H. EUGENE, ed. *Biomedical Physics and Biomaterials Science.* Cambridge, Mass.: MIT Press, 1972.

STODDART, ANNA M. *The Life of Paracelsus, Theophrastus von Hohenheim 1493–1541.* London: 1915.

THOMPSON, C. J. S. *The History and Evolution of Surgical Instruments.* New York: Schuman's, 1942.

———. *The Mystery and Romance of Alchemy and Pharmacy.* London: The Scientific Press, 1897.

"Trace Elements Take Spotlight at Giant Biology Conference." *Medical World News* XIV (11 May 1973): 36–37.

TREASE, GEORGE EDWARD. *Pharmacy in History.* London: Baillière, Tindall and Cox, 1964.

WALLACE, DAVID M. "New Lamps for Old," *Proceedings of the Royal Society of Medicine* LXVI (May 1973): 455–58.

WELCH, TERESA J. C. et al. *Fundamentals of the Tracer Method.* Philadelphia: Saunders, 1972.

WILLIAMS, D. F. and ROAF, ROBERT. *Implants in Surgery.* Philadelphia: Saunders, 1973.

WILSON, WILLIAM JEROME. "The Origin and Development of Greco-Egyptian Alchemy." *Ciba Symposia* III, no. 5 (August 1941): 925–60.

WOOTTON, A. C. *Chronicles of Pharmacy.* 2 vols. London: Macmillan, 1910.

Illustration Credits

Page xiii: John Stewart Milne, *Surgical Instruments in Greek and Roman Times* (Oxford: Clarendon Press, 1907), plate 4. Courtesy of the publisher

Page xiv: A. C. Wootton, *Chronicles of Pharmacy* (London: Macmillan Co., 1910), Vol. 1, page 225. Courtesy of the publisher

Page xv: Northwestern University Medical Library Portrait File

Pages xvii and xviii: Ambroise Paré, *Dix Livres de la Chirurgie avec le Magasin des Instrumens necessaires a icelle* (Paris, 1564), leaves 123 and 226, respectively

Page xix: Samuel Sharp, *A Treatise on the Operations of Surgery* (London: J. Watts, 1739), plate 9

Page xx, top: J. A. Chalmers, "James Young Simpson and the 'Suction-Tractor,' " *Journal of Obstetrics and Gynaecology of the British Commonwealth* 70 (1963):96, fig. 4. Courtesy of the author and the editor

Page xx, bottom: T. G. H. Drake, "Antique Pewter Articles of Medical Interest," *Bulletin of the History of Medicine* 10 (1941):276. Courtesy of the editor

Page xxi: Copy of cartoon by James Gillray, Northwestern University Medical Library Portrait File

Page xxii: Gilbert Barling, "The Electric Cystoscope and the Method of Using It," *Birmingham Medical Review* 25 (1889):258

Page xxiii: Howard A. Kelly, "The Cystoscope," *American Journal of Obstetrics* 30 (1894):86 and 90

Page xxiv: Thomas Annandale, "On the Use of Steel Pins in the Practice of Surgery," *Scottish Medical and Surgical Journal* 1 (1897):869

Page xxv, top: Courtesy of Baird Atomic, Bedford, Mass.

Page xxv, bottom: Courtesy of G. D. Searle, Chicago, Ill.

Index

INDEX